Russian and Ukrainian Perspectives on the Concept of Information Confrontation

Translations, 2002–2020

MICHELLE GRISÉ, YULIYA SHOKH, KHRYSTYNA HOLYNSKA, ALYSSA DEMUS

Prepared for the U.S. European Command
Approved for public release; distribution unlimited

For more information on this publication, visit **www.rand.org/t/RRA198-7**.

Published by the RAND Corporation, Santa Monica, Calif.

Library of Congress Cataloging-in-Publication Data is available for this publication.

ISBN: 978-1-9774-0777-1

Cover Images:
Kremlin: skyNext/Getty Images/iStockphoto.
Circuit Board: DKosig/Getty Images/iStockphoto.

Cover Designer: Rick Penn-Kraus.

About This Report

The role of information and information technologies in strategic competition and military operations has evolved considerably in the first two decades of the 21st century. The Russian and Ukrainian authors of the translated articles in this volume, which were published between 2002 and 2020, provide insight into the evolution of military-scientific thinking in both Russia and Ukraine on the concept of information confrontation (*informatsionnoe protivoborstvo*). This volume will be useful for researchers and policymakers who are interested in understanding how the Russian and Ukrainian military-scientific communities are debating the increasing prominence of information confrontation and information warfare in periods of strategic competition and times of conflict. Translations were produced with the permission of East View Press, M. M. Prysiazhniuk, Irina Yuzova, and the *Anthology of Research Works of Kharkiv National Air Force University.*

The research reported here was completed in March 2021 and underwent security review with the sponsor and the Defense Office of Prepublication and Security Review before public release.

RAND National Security Research Division

This work was sponsored by the United States European Command and conducted within the International Security and Defense Policy Center of the RAND National Security Research Division (NSRD), which operated the RAND National Defense Research Institute (NDRI), a federally funded research and development center (FFRDC) sponsored by the Office of the Secretary of Defense, the Joint Staff, the Unified Combatant Commands, the Navy, the Marine Corps, the defense agencies, and the defense intelligence enterprise.

For more information on the RAND International Security and Defense Policy Center, see www.rand.org/nsrd/isdp or contact the director (contact information is provided on the webpage).

Acknowledgments

We would like to acknowledge the support of our sponsor, Ken Stolworthy, Director of the Russia Strategic Initiative, U.S. European Command. Within our sponsor's office, we would like to thank Col. Phil Forbes, United States Air Force, for his support and assistance.

We would also like to acknowledge the RAND colleagues who offered their expertise, support, and guidance throughout this effort. In particular, we would like to acknowledge and thank those responsible for leading this effort, including Mark Cozad, Dara Massicot, and Clint Reach. We also thank Elina Treyger and Chris Paul, who provided helpful feedback on an earlier draft of this report. Finally, we would like to thank Christine Wormuth, Agnes Schaefer, and Francisco Walter for helping us make this document come to fruition.

Contents

Figures and Tables

Figures

Tables

Summary

Issue

The role of information and information technologies in strategic competition and military operations has evolved considerably, both in complexity and prominence, in the first two decades of the 21st century. The authors of the translated articles in this volume, which were published between 2002 and 2020, provide insight into the evolution of military-scientific thinking in both Russia and Ukraine on the concept of *information confrontation* (*informatsionnoye protivoborstvo*), defined by one author as the "purposeful use of information to achieve political, economic, military, and other goals."[1]

The articles in this volume detail the impact of the rapid development of information technologies and information weapons in recent years on the military-scientific literature of Russia and Ukraine. Although the authors of the earlier articles in this volume, V. Slipchenko and V. I. Orlansky, acknowledge the growing importance of information in warfare, the authors of the articles published since 2012 show the rapid informatization of all aspects of strategic competition and military operations and call for the development of a unified system and comprehensive strategy for information confrontation.

Approach

This volume includes articles representing nearly two decades of commentary on the concept of information confrontation: This illustrates how leading members of the military-scientific communities in both Russia and

---

[1] K. I. Sayfetdinov, "Informatsionnoye protivoborstvo v voyennoy sfere [Information Confrontation in the Military Sphere]," *Voennaya mysl' [Military Thought]*, No. 7, 2014, p. 38.

Ukraine have wrestled with and debated the roles of information weapons and information influence in periods of strategic competition and times of conflict. The primary factor in deciding which articles to include in this volume was the expertise and stature of the authors.[2] Although the concept of information confrontation has received significant attention in military journals and more-popular forums, the authors that we included in this volume have conducted research on information confrontation and related concepts over a period of decades and established themselves as leading experts in the field. Major General Vladimir Slipchenko, for example, wrote on information confrontation for almost 20 years. In this volume, we include two articles authored by Slipchenko in 2002 and 2013 to illustrate how Russian thinking on the evolving role of information confrontation in the modern era has changed to reflect advancements in information technologies and the increasingly central role of information in conflict. The other authors included in this volume are affiliated with some of the leading military academies and universities in Russia and Ukraine. The professional stature and credibility of the selected authors is further reinforced by the publication of their work in *Military Thought* (*Voennaya mysl'*), Russia's most prestigious military journal, as well as in other leading military journals, including the *Journal of the Academy of Military Sciences* (*Vestnik Akademii voennykh nauk*), and *Army Digest* (*Armeiskii sbornik*).

A secondary factor in the article selection was the scope and diversity of substantive insights offered. The articles in this volume provide insight into

[2] To select the articles included in this volume, we first identified the major military-scientific journals that have published articles on information operations and information warfare. We conducted a search for the Russian term *informatsionnoe protivoborstvo* (information confrontation) within those journals, although we also searched for related Russian terms, such as *informatsionnaya voyna* (information war), to understand the difference between these related terms. We then compiled a list of articles that satisfied two criteria: First, as noted previously, their authors have significant experience in the field and have frequently published on related topics; and second, the articles focused primarily on the concept of information confrontation and provided a comprehensive and high-quality analysis of its role in Russian strategic thinking. From this list of potential articles, we selected the articles that are included in this volume because they provide readers with a variety of viewpoints on information confrontation, including both the strategic significance of the concept and its operational and tactical implications.

the varying definitions and subtypes of information confrontation, its historical evolution and application, the technical tools used in the conduct of information confrontation, and the relationship between the connected yet distinct concepts of information confrontation and information warfare. In addition to providing insight into Russian concepts of information confrontation, this volume includes several Ukrainian works, including a chapter from an edited volume and a journal article, that also focus on information confrontation and the conduct of information warfare but additionally consider effective means of defending against information confrontation. These works were chosen based on the expertise and professional stature of their authors within the Ukrainian military-scientific community, and were selected to illustrate the effect of the application of principles of information confrontation by Russia in Ukraine in recent years on Ukrainian understandings of this concept.

List of Authors

Name	Rank/Position	Education
V. A. Annenkov	Colonel (ret.); senior researcher, Research Center of Peter the Great Military Academy of the Strategic Rocket Troops	Ph.D. in military sciences
S. V. Golubchikov	Lead researcher, Almaz Research and Development Company	Ph.D. in technical sciences
V. F. Lata	Lieutenant General (ret.); principal researcher, Research Center of Peter the Great Military Academy of the Strategic Rocket Troops	Ph.D. in military sciences
V. F. Moiseev	Colonel (ret.); lead researcher, Research Center of Peter the Great Military Academy of the Strategic Rocket Troops	Ph.D. in technical sciences
V. K. Novikov	Major General (ret.); professor, Peter the Great Military Academy of the Strategic Rocket Troops	Ph.D. in technical sciences
V. I. Orlansky	Colonel; professor, Military Academy of the General Staff of the Russian Armed Forces	Ph.D. in military sciences
M. M. Prysiazhyuk	Associate professor, Military Institute of Taras Shevchenko National University of Kyiv	Ph.D. in technical sciences
K. I. Sayfetdinov	Major General (ret.)	Ph.D. in military sciences
V. Slipchenko	Major General (ret.)	Ph.D. in military sciences
K. A. Trotsenko	Colonel; Commander, Southern Military District	Ph.D. in military sciences
I. Yuzova	Acting chief, Sergeants' College, Ivan Kozhedub National Air Force University, Kharkiv, Ukraine	Ph.D. in technical sciences

CHAPTER ONE

Introduction

The role of information and information technologies in strategic competition and military operations has evolved considerably, both in complexity and prominence, in the first two decades of the 21st century. The authors of the translated articles in this volume, which were published between 2002 and 2020, provide insight into the evolution of military-scientific thinking in both Russia and Ukraine on the concept of *information confrontation* [*informatsionnoe protivoborstvo*], defined by one author as the "purposeful use of information to achieve political, economic, military, and other goals."[1]

The articles in this volume detail the impact of the rapid development of information technologies and information weapons in recent years on the military-scientific literature of Russia and Ukraine. Although the authors of the earlier articles in this volume (V. Slipchenko and V. I. Orlansky) acknowledge the growing importance of information in warfare, the articles that were published since 2012 reflect the rapid informatization of all aspects of strategic competition and military operations; these articles call for the development of a unified system and comprehensive strategy for information confrontation.

The articles in this volume also refer to several terms and concepts that often are used in conjunction with the study and analysis of information confrontation. Although some of these terms, such as *informatization*, are used to describe foundational elements of information confrontation, others, such as *information warfare*, are frequently used synonymously

---

1 K. I. Sayfetdinov, "Informatsionnoye protivoborstvo v voyennoy sfere [Information Confrontation in the Military Sphere]," *Voennaya mysl' [Military Thought]*, No. 7, 2014, p. 38.

with information confrontation, and still others, such as *information war*, appear as the subject of great debate in the literature on information confrontation. Nonetheless, our analysis of the literature see suggests that these related terms are distinct from the concept of information confrontation and have their own differential meanings.[2] Table 1.1 lists and defines the related terms used in this volume.

There are two related concepts discussed throughout the literature: information confrontation and information warfare. *Information confrontation*, as K. I. Sayfetdinov notes, is a task that should be conducted "constantly" in peacetime.[3] It is a broader concept, understood as a "multifaceted, multifactorial"[4] struggle that encompasses "social systems, classes, nations, [and] states through diplomatic, political, informational, psychological, financial, economic influence, armed conflict, and many other forms . . . to achieve strategic and political goals,"[5] as V. Slipchenko explains, while *information warfare* consists of information operations during active conflict. Although information warfare is the primary focus during wartime, techniques of information confrontation still play important roles during military conflict through "gain[ing] and maintain[ing] information superiority over the enemy's armed forces" while simultaneously "creat[ing] favorable conditions for the preparation and use of [Russia's] armed forces."[6] In the modern era, information confrontation is conducted at all points along the

---

2 For a comprehensive analysis of these terms and their relationship to the concept of information confrontation, see the companion report to this volume, Michelle Grisé, Alyssa Demus, Yuliya Shokh, Marta Kepe, Jonathan Welburn, and Khrystyna Holynska, *Rivalry in the Information Sphere: Russian Conceptions of Information Confrontation*, Santa Monica, Calif.: RAND Corporation, RR-A198-8, 2022.

3 Sayfetdinov, 2014, p. 39.

4 V. Slipchenko, "Informatsionnyy resurs i informatsionnoye protivoborstvo [Information Resources and Information Confrontation]," *Armeiskii sbornik [Army Digest]*, No. 10, 2013, p. 54.

5 Slipchenko, 2013, p. 53.

6 Sayfetdinov, 2014, p. 39.

TABLE 1.1

Main Terms Related to Information Confrontation Used in This Volume

Related Term[a]	Definition
Informatization	Phenomenon that makes it possible to engage in information confrontation; serves as the foundation for the growing role of information activities and operations and their impact on modern society[b]
Information war	This term is defined in the following ways: • "Confrontation between two or more states in the information domain with the purpose of causing damage to information systems, processes and resources, critical and other infrastructure, undermining the political, economic and social systems, massive psychological manipulation of the population to destabilize the state and society, as well as coercing the state to make decisions in the interest of the opposing force"[c] • "Transparent and severe clash between states" characterized by causing "harmful impact on the information domain"[d] • Struggle between opposing sides for superiority over the enemy in timeliness, assurance, completeness of information, speed and quality of its processing and dissemination[e] • Use of "aggressive information influence"[f]
Information warfare	Activities undertaken to gain information superiority in the process of armed confrontation[e,g,h]
Information operations	Set of information activities that are coordinated in terms of purpose, objects, place, and time, and are conducted to gain and maintain information superiority over the enemy or reduce the enemy's information superiority in a given combat theater or strategic direction[g,i]
Hybrid wars	Not defined in Russian strategic documents; however, Russian military leaders and scholars have observed the following: • Origins of hybrid or "multimodal" wars can be traced to U.S. and NATO aggression in the former Yugoslavia[j] • "Hybrid war requires high-tech weapons and scientific justification" to support the use of minimal armed forces against the enemy[k] • Military actions that combine military, diplomatic, information, economic, and other means to achieve strategic policy goals[l]

Table 1.1—Continued

[a] Although some articles included in this volume reference these terms, those articles might not provide a clear or comprehensive definition of these terms. To augment the definitions, we have cited other sources as well.
[b] V. S. Shevtsov, "Informatsionnoye protivoborstvo v globaliziruyushemsia mire: Aktual'nost', differentsiatsiya poniatiy, ugrozy politicheskoy stabil'nosti [Information Confrontation in a Globalizing World: Relevance, Differentiation of Concepts, Threats to Political Stability]," *University Bulletin [Vestnik Universiteta]*, No. 5, 2015.
[c] Ministry of Defense of the Russian Federation, *Kontseptual'nye vzglyady na deyatel'nost' Vooruzhennykh sil Rossiyskoy Federatsii v informatsionnom prostranstve [Conceptual Views on the Activities of the Armed Forces of the Russian Federation in the Information Space]*, 2011.
[d] Ministry of Defense of the Russian Federation, "Informatsionnaya voyna [Information War]," *Voyennyy entsiklopedicheskiy slovar' [Military Encyclopedic Dictionary]*, trans. Polina Kats-Kariyanakatte, Joe Cheravitch, and Clint Reach, webpage, undated-a.
[e] Y. Nuzhdin, "Informatsionniye voyny. Uroki devianostykh [Information Wars. Lessons of the Nineties]," *Flag Rodiny [Flag of the Motherland]*, November 22, 2000.
[f] Ministry of Defense of the Russian Federation, "Ministr oborony Sergey Shoygu nazval glavnoy tsel'yu informatsionnoy voyny Zapada protiv Rossii polnoye yemu podchineniye [Defense Minister Sergey Shoygu Called Complete Submission to the West as the Main Goal of the Information War of the West Against Russia]," webpage, June 26, 2019.
[g] M. A. Rodionov, "K voprosu o formakh vedeniya informatsionnoy bor'by [On the Question of the Ways of Waging Information Warfare]," *Voennaya mysl' [Military Thought]*, No. 2, 1998.
[h] V. Slipchenko, "Novaya forma bor'by. V nastupivsheme veke rol' informatsii v beskontaktnykh voynakh budet lish' vozrastat' [A New Form of Combat. In the Coming Century the Role of Information in the Contactless Wars Will Only Increase]," *Armeiskii sbornik [Army Digest]*, No. 12, 2002.
[i] M. Prysiazhniuk, "Osoblyvosti suchasnoho periodu informatsiyno-psykholohichnoho protyborstva [Peculiarities of the Modern Period of Informational-Psychological Confrontation]," in Y. Zharkov et al., eds., *Istoriia informatsiino-psykholohichnoho protyborstva [History of Information and Psychological Confrontation]*, Kyiv, Ukraine: Research and Publishing Department of the National Academy of Security Service of Ukraine, 2012.
[j] Yu. Matvienko, "'Tsvetniye' revoliutsii kak nevoyenniy sposob dostizheniya politicheskikh tseley v gibridnoy voyne: Sushnost', soderzhaniye, vozmozhniye mery zashity i protivodeystviya ['Color' Revolutions as Non-Military Means to Achieve Political Goals in 'Hybrid' War: Nature, Content, Possible Protection Measures and Countermeasures]," *Informatsionniye Voyny [Information Wars Journal]*, Vol. 4, No. 40, 2016.
[k] Valery Gerasimov, "Po opytu Sirii [Syrian Experience]," *Voyenno-Promyshlennyy Kur'er Online [Military-Industrial Courier Online]*, March 7, 2016.
[l] I. Yuzova, "Analiz Orhanizatsiyi ta vedennya informatsiyno-psykholohichnykh operatsiy pry vedenni hibrydnoyi viyny [Analysis of the Organization and Conduct of Informational-Psychological Operations in the Conduct of Hybrid Warfare]," *Zbirnyk naukovykh prats' Kharkivs'koho natsional'noho universytetu Povitryanykh Syl [Anthology of Research Works of Kharkiv National Air Force University]*, No. 2, 2020.

continuum, from peacetime to wartime. As V. I. Orlansky notes, in today's world, "everything has become informational."[7]

Russian military scholars have identified two main subtypes of information confrontation, informational-psychological and informational-technical. The informational-psychological aspect of information confrontation includes efforts to influence the enemy's population and military forces,[8] including by "mislead[ing] the enemy, undermin[ing] its will to resist, produc[ing] panic in its ranks, and generat[ing] betrayal."[9] The informational-technical component of information confrontation, on the other hand, involves the physical manipulation of information networks and tools, including the "destruction of information, radio-electronic, [and] computer networks, and [gaining] unauthorized access to the information resources of the enemy."[10] Although this typology is used consistently throughout this translation volume, and in the military-scientific literature more broadly, a "unified system of terms, concepts, and definitions" related to information confrontation remains elusive.[11]

Summary of Articles

In the first article in this volume, V. F. Lata, V. A. Annenkov, and V. F. Moiseev note that information "has been the target of warfare" throughout history.[12] They assert, however, that the rise of the information age has led to a

---

[7] V. I. Orlansky, "Informatsionnoye oruzhiye i informatsionnaya bor'ba: Real'nost' i domysly [Information Weapons and Information Warfare: Reality and Speculation]," *Voennaya mysl' [Military Thought]*, No. 1, 2008, p. 62.

[8] K. A. Trotsenko, "Informatsionnoye protivoborstvo v operativno-takticheskom zvene upravleniya [Information Confrontation on the Operational-Tactical Level]," *Voennaya mysl' [Military Thought]*, No. 8, 2016, p. 20.

[9] Sayfetdinov, 2014, p. 38.

[10] Trotsenko, 2016, p. 20; Orlansky, 2008, p. 66.

[11] V. F. Lata, V. A. Annenkov, and V. F. Moiseev, "Informatsionnoye protivoborstvo: Sistema terminov i opredeleniy [Information Confrontation: System of Terms and Definitions]," *Vestnik Akademii Voennykh Nauk [Journal of the Academy of Military Sciences]*, No. 2, 2019, p. 129.

[12] Lata, Annenkov, and Moiseev, 2019, p. 128.

new kind of constant information struggle, both between states and within states. As a result of this shift, there has been a proliferation of interrelated yet distinct terms in the military-scientific literature, such as information war, information weapon, information resource, information space, information domain, and information security. Scholars have defined these terms differently, however, which has complicated the development of a "unified understanding" of the role of information in modern warfare.[13] The authors of this article propose the development of an "internally consistent," "single system of terms, concepts, and definitions" related to information confrontation.[14] By examining the distinct features of key terms, they provide a starting point for the development of this unified system.

V. K. Novikov and S. V. Golubchikov characterize the history of warfare as a history of technological development. They write: "From one war to the next war, there [has] been a continuous process of improving the destructive factors of weapons [and] the means of their delivery."[15] The authors suggest that information confrontation and information weapons are tools that represent only the most recent stage of this historical trajectory. Similar to several other authors in this volume, Novikov and Golubchikov argue that the Persian Gulf War marked the beginning of modern information warfare.[16] They provide a detailed chronology of information wars (defined in Table 1.1) since the end of the Cold War, noting that since the early 1990s, the United States and the West more broadly have conducted information wars 39 times. At the conclusion of the article, the authors make a series of recommendations for how Russia can better adapt to an age of informatized warfare, including the amendment of existing Russian legislation and additional investment in mathematics and science education for Russian students.[17]

---

[13] Lata, Annenkov, and Moiseev, 2019, p. 129.

[14] Lata, Annenkov, and Moiseev, 2019, p. 129.

[15] V. K. Novikov and S. V. Golubchikov, "Analiz informatsionnykh voyn za posledniye chetvert' veka [Analysis of Information Wars of the Last Quarter Century]," *Vestnik Akademii Voennykh Nauk [Journal of the Academy of Military Sciences]*, No. 3, 2017, p. 10.

[16] Novikov and Golubchikov, 2017, pp. 11–12.

[17] Novikov and Golubchikov, 2017, p. 16.

In another example, Slipchenko traces the origins of information confrontation to the Persian Gulf War. In his 2002 article in which he writes about the role of information confrontation in "contactless wars," he argues that information serves as an adjunct to other methods of warfare.[18] He predicts that, by the middle of the 21st century, information will "acquire an independent character."[19] Until then, Slipchenko argues that there will not be a true information war but rather information confrontation.[20] In his 2013 article, Slipchenko acknowledges that the technological development of information weapons in the West had occurred faster than he had originally anticipated and reflects on the future of information confrontation and information warfare. Although his analysis focuses on information weapons, he also predicts that 2050 will be a watershed moment for the military applications of artificial intelligence. He notes that in the first decade of the 21st century, states made significant progress in developing the technologies necessary for achieving information superiority over their adversaries.[21]

Writing in 2008, V. I. Orlansky, like Slipchenko in his earlier work, characterizes information technologies and information weapons as playing supporting roles in warfare. "Information has always played a supporting role," he writes, and "for all the importance of information, it does not replace, and possibly will never replace, weapons, [and] it will not become the main means of conducting [warfare]."[22] Orlansky concludes that information has "not yet become a means that is comparable in terms of the strength of [its] impact to traditional weapons."[23] His perception of the supporting role of information in warfare can be contrasted with the views expressed in more-recent articles featured in this volume, which reflect the growing capabilities of information technologies and, as a result, the increasingly central role of information as a means of engaging in both strategic competition and warfare.

[18] Slipchenko, 2002, p. 30.

[19] Slipchenko, 2002, pp. 30–31.

[20] Slipchenko, 2002, p. 31.

[21] Slipchenko, 2013, p. 54.

[22] Orlansky, 2008, p. 63.

[23] Orlansky, 2008, p. 63.

K. I. Sayfetdinov acknowledges that the primary goal of information confrontation, at least in the military sphere, is to "achieve and maintain information superiority over the enemy's armed forces."[24] He also devotes significant attention to its role in peacetime and during competition in the article translated in this volume. In peacetime, he explains, information confrontation serves as a tool of strategic deterrence that allows the political leadership of Russia to more effectively promote national security through a variety of political, diplomatic, economic, legal, and military measures.[25] During competition, information confrontation becomes a way for the military-political leadership of the country to solve specific problems and achieve its goals.[26] Like Lata, Annenkov, and Moiseev, Sayfetdinov emphasizes the importance of conceiving of information confrontation as an integrated system. As he sees it, the "system of information confrontation" comprises a number of subsystems, including information technology, intelligence, electronic warfare, and psychological struggle.[27] This system is depicted in Figure 1.1. Sayfetdinov recommends the construction of an integrated system of information confrontation that takes into account the specific function of each of these subsystems.[28]

Unlike the other authors represented in this volume, K. A. Trotsenko focuses on the role of information confrontation at the operational and tactical levels. He identifies two primary means of achieving information superiority: (1) through a direct impact on the flow of information, and (2) by affecting the "critical nodes of the enemy's armed warfare processes."[29] He examines the implications of each for command and control. Trotsenko suggests that military strategists can learn from the successes and failures

[24] Sayfetdinov, 2014, p. 39. According to this view, information superiority over an adversary's forces is critical both during preparations for conflict and during conflict itself.

[25] Sayfetdinov, 2014, p. 40.

[26] Sayfetdinov, 2014, p. 40.

[27] Sayfetdinov, 2014, p. 41.

[28] Sayfetdinov, 2014, p. 41.

[29] Trotsenko, 2016, p. 22.

FIGURE 1.1
Integrated System of Information Confrontation

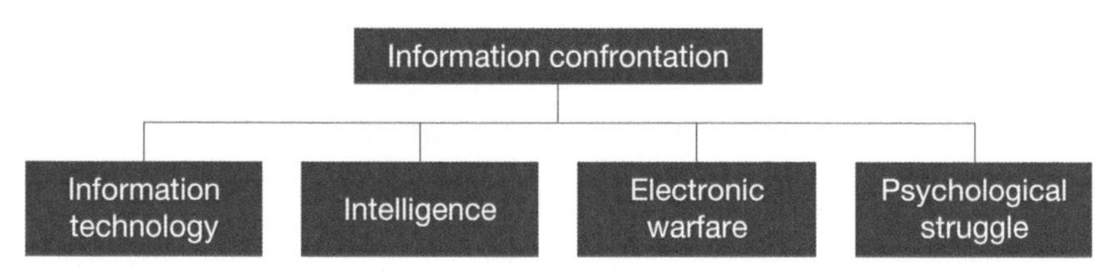

SOURCE: Adapted and translated from Sayfetdinov, 2014, p. 41.

of conventional forces in past conflicts as they consider the practical implementation of the concept of information confrontation.

Like the Russian scholars featured in this volume, Ukrainian scholar M. Prysiazhniuk characterizes information confrontation as having ancient roots. He notes that even in ancient times, soldiers used informational tools to influence their enemies.[30] Prysiazhniuk examines the history of information confrontation and categorizes different types of information confrontation. By contrast, another Ukrainian scholar, I. Yuzova, examines the practical application of information confrontation in hybrid wars (defined in Table 1.1), placing an emphasis on the informational-psychological type of information confrontation. Yuzova notes that the "reality of what is taking place in Ukraine today," and, in particular, the "colossal informational, psychological, and military support" to the conflict from Russia underscores the critical necessity of developing ways to defend against informational-psychological influence.[31]

Selection of Articles

By including articles that represent nearly two decades of commentary on the concept of information confrontation, this volume illustrates how leading members of the military-scientific communities in both Russia and Ukraine have wrestled with and debated the role of information weapons

---

[30] Prysiazhniuk, 2012, p. 140.

[31] Yuzova, 2020, p. 42.

and information influence in periods of strategic competition and times of conflict. The primary factor in deciding which articles to include in this volume was the expertise and stature of the authors.[32] Although the concept of information confrontation has received significant attention in military journals and more-popular forums, the authors featured in this volume have conducted research on information confrontation and related concepts over a period of decades and established themselves as leading experts in the field. Slipchenko, for example, wrote on information confrontation for almost 20 years. In this volume, we include two articles authored by Slipchenko, in 2002 and 2013, respectively, to illustrate how Russian thinking and perceptions of the evolving role of information confrontation in the modern era have changed to reflect advancements in information technologies and the increasingly central role of information in conflict. The other authors included in this volume are affiliated with some of the leading military academies and universities in Russia and Ukraine. The professional stature and credibility of the selected authors is further reinforced by the publication of their work in *Military Thought* (*Voennaya mysl'*), Russia's most prestigious military journal, and other leading military journals, including the *Journal of the Academy of Military Sciences* (*Vestnik Akademii voennykh nauk*), and *Army Digest* (*Armeiskii sbornik*).

A secondary factor in the article selection was the scope and diversity of substantive insights offered. The articles in this volume provide insight into the varying definitions and subtypes of information confrontation, its

[32] To select the articles included in this volume, we first identified the major military-scientific journals that have published articles on information operations and information warfare. We conducted a search for the Russian term *informatsionnoe protivoborstvo* (information confrontation) within those journals, although we also searched for related Russian terms, such as *informatsionnaya voyna* (information war) to understand the difference between these related terms. We then compiled a list of articles that satisfied two criteria: First, as noted previously, their authors have significant experience in the field and have frequently published on related topics; second, the articles are focused primarily on the concept of information confrontation and provide a comprehensive and high-quality analysis of its role in Russian strategic thinking. From this list of potential articles, we selected the articles that are included in this volume because they provide readers with a variety of viewpoints on information confrontation, including both the strategic significance of the concept and its operational and tactical implications.

historical evolution and application, the technical tools used in its conduct, and the relationship between the connected yet distinct concepts of information confrontation and information warfare. In addition to providing insight into Russian conceptions of information confrontation, this volume includes several translated Ukrainian works, including a chapter from an edited volume and a journal article, which also focus on information confrontation and the conduct of information warfare but additionally consider effective means of defending against information confrontation. These works were chosen based on the expertise and professional stature of their authors within the Ukrainian military-scientific community. They were selected to illustrate the effect of the Russian application of principles of information confrontation in Ukraine in recent years on Ukrainian understandings of this concept.

Overview of Articles

In the following list, we provide a brief overview of each of the articles translated in this volume:

- **V. Slipchenko, "A New Form of Combat," 2002**. In this article, the author provides insight into the concept of information confrontation at the beginning of the 20th century. The author characterizes information confrontation as adjunct to other methods of warfare.
- **V. I. Orlansky, "Information Weapons and Information Warfare: Reality and Speculation," 2008**. The author of this article characterizes information weapons as playing supporting roles in warfare, having not yet become comparable with traditional weapons.
- **M. Prysiazhniuk, "Peculiarities of the Modern Period of Informational-Psychological Confrontation," 2012**. In this chapter, which is from an edited volume on informational-psychological confrontation, the author discusses the historical antecedents of information confrontation, noting that informational tools have been used since ancient times to influence one's adversaries.
- **V. Slipchenko, "Information Resources and Information Confrontation," 2013**. In this article, the author notes that the technological

development of information weapons has proceeded rapidly and predicts that information confrontation will play a central role in future conflicts.

- **K. I. Sayfetdinov, "Information Confrontation in the Military Sphere," 2014.** The author explains the role of information confrontation in peacetime, during competition, and in wartime. He then proposes the development of an integrated system of information confrontation.
- **K. A. Trotsenko, "Information Confrontation on the Operational-Tactical Level," 2016.** The author examines the role of information confrontation at the operational and tactical levels.
- **V. K. Novikov and S. V. Golubchikov, "Analysis of Information Wars of the Last Quarter Century," 2017.** In this article, the authors provide a broad examination of the history of warfare—a history that has been driven by the introduction of new technological innovations. They characterize information confrontation as representing the most recent stage of this historical trajectory.
- **V. F. Lata, V. A. Annenkov, and V. F. Moiseev, "Information Confrontation: System of Terms and Definitions," 2019.** The authors of this article examine the definitions of information confrontation and related terms. They propose the development of a unified system of terms related to information confrontation.
- **I. Yuzova, "Analysis of the Organization and Conduct of Informational-Psychological Operations in the Conduct of Hybrid Wars," 2020.** The author of this article examines the practical application of information confrontation in hybrid wars, including the conflict in Ukraine.

CHAPTER TWO

A New Form of Combat: In the Coming Century, the Role of Information in Contactless Wars Will Only Increase

The rise in interest in information confrontation in the wars of the future is not an accident. This is linked to the fact that information is becoming the same type of weapon as rockets, torpedoes, etc. Today, it is clear that information confrontation will become a factor that will have a significant impact on the beginning, conduct, and result of the wars of the future.[1]

One of the most important mechanisms for the emergence of contactless wars is not only the revolution in military affairs that is now taking place in developed countries but also the information-scientific and technological revolution that is also forming completely new information systems on a global scale. Of course, during the transition period to contactless wars, until about 2010, many elements of the confrontation of the past generation (contact wars) will remain. However, it is already clear that a sharp leap toward informatization and automation of command and control of troops and weapons is brewing. Here, we should also expect a rapid process of automation of all levels of the organizational structure of the armed forces. But during the transition, information confrontation will remain only one of the types of support for other methods of warfare.

Nevertheless, we can foresee that, during the transition to contactless wars, [information confrontation] will gradually go beyond the support type

---

[1] [Slipchenko, 2002.]

and become combat; that is, it will acquire an independent character among many other forms and methods of warfare. Superiority over the enemy will be achieved through obtaining an advantage in diverse types of information, mobility, reaction speed, and in [having] accurate fires and information effects[2] in real time against numerous targets of the enemy's economy and defense with the lowest possible risk of damage to one's own forces and means. However, unlike high-precision shock weapons that strike a specific, specially selected important target or its critical node, an information weapon will be system-destructive; that is, [it will] incapacitat[e] entire combat, economic, or social systems.

The possession of information resources in the wars of the future will become the same indispensable attribute as before—the possession of forces and means, weapons and ammunition, transport, etc. Winning information confrontation during contactless wars can help achieve strategic goals.

Thus, information confrontation in contactless wars should be understood as a new, strategic form of warfare between the parties, in which special methods and means are used that deliver effects against the enemy's information environment and protect one's own [information environment] in the interests of achieving the strategic goals of the war. However, as a form of support for military operations, the struggle for the possession of information is inherent in almost all past generations of wars. It is going on now, since the opposing sides always strive to control the enemy's information accordingly not only in wartime, but also in peacetime.

In its most general form, the main goal of information confrontation is to maintain the required level of one's own information security and reduce the level of such security for the enemy. This two-pronged task can be achieved with the help of combined effects that are aimed at destroying the enemy's information resources and domain while maintaining one's own.

It is quite obvious that, for confrontation in contactless wars, the information resource of high-precision weapons must have a full set of software tools for both active and passive protection measures. It will be necessary not only to protect the information systems of high-precision weapons from attacks, but also to carry out active and passive effects against all existing

2 [*Information effects* refer to the technical effects of information operations.]

and future enemy air defense and missile defense systems. It is likely that information confrontation will be closely linked with intelligence systems and means.

Such experience was "acquired" by the world community for the first time in 1991, during the contactless war in the Persian Gulf. At that time, the multinational force, conducting a specially planned radio-electronic and fires counteraction on an operational scale, successfully blocked practically the entire state and military information system of Iraq. The second "experiment" was tested in 1999 in [the former] Yugoslavia. The successes achieved not only inspired the United States, multinational forces, and NATO [North Atlantic Treaty Organization] countries, who realized the role of information confrontation in contactless wars, but also made those entities think about how to deal with a situation in which they face the same type of confrontation.

However, some domestic and foreign scholars believe that information war, not information confrontation, is already being waged and will continue to be waged. But the concept of "war" in this context is generally inappropriate because it refers to a more complex sociopolitical phenomenon—a specific state of society associated with a sharp change in relations between states, peoples, social groups and characterized by the use of armed violence to achieve political, economic, and other purposes. Moreover, war, in its classical understanding, is not only information confrontation, but also a confrontation between social systems, classes, nations, [and] states using diplomatic, political, informational, psychological, financial, economic impacts, armed forces, and many other forms and methods of warfare to achieve strategic and political goals.

I think that, at least in the next 20–40 years, we should not yet expect the next generation of wars that are contactless but constitute information wars. If such wars arise in the future, they will certainly be fought in the global information space and mainly by informational means. However, this next, seventh generation of wars could arise no earlier than 50 years from now. But until then, there will be information confrontation, not war.

We can assume that wars of the next generation will inevitably go beyond the operational and even strategic scale and immediately become global. Moreover, it is already obvious that they will not be conducted using only informational means. Although, using [informational means], the aggressor will be able to provoke man-made disasters in large economic areas, regions, and parts of the world. It is possible that, after 2050, in the course of information wars, economic weapons could be developed to target a country's mineral and biological resources at separate, localized areas of the biosphere (atmosphere, hydrosphere, lithosphere) and the climate resources of the earth.

However, the choice of the listed "targets" clearly exposes the absurdity of waging such a war because it will be associated not only with the processes of disrupting the normal operations of the global information space and resources but also the environment supporting all life on earth. It is also fundamentally important that an individual will not be the main target of defeat in the wars of subsequent generations (starting from the sixth—contactless wars) because defeating other structures and systems that sustain life will indirectly affect the individual.

As for contactless wars (sixth generation), "information confrontation" and "information warfare" are completely legitimate concepts and express the struggle between opposing sides for superiority over the quantity, quality, and speed of collecting, analyzing, and using information. It is clear that this type of confrontation, like its other types in contactless wars, already has two clearly defined components: defensive and offensive (shock).

Defensive [confrontation] consists of protecting one's own information infrastructure and information itself from the enemy and ensuring the security of one's own information resources.

Offensive [confrontation] consists of disrupting or destroying the enemy's information infrastructure and disrupting their process of operational control over their forces and means.

Such tools and methods of ensuring the security of one's own information systems and resources as operational and strategic camouflage, physical protection of information infrastructure facilities, disinformation, electronic warfare, and others can be considered defensive factors in contactless wars.

Such methods of warfare as strategic camouflage, disinformation, electronic warfare, physical destruction and annihilation of information infrastructure targets, "attacks" on the enemy's computer networks, "information effects," "information intrusion" or "information aggression," and "information strikes" can be employed as offensive factors.

It is possible that cyber warfare can develop independently within the information confrontation framework, during which powerful information strikes would be delivered against the enemy's integrated computer systems. Information intrusion can be carried out through the internet to disrupt the enemy's life-sustaining systems, communications, electricity, gas, and water supplies, paralyze traffic, disrupt financial transactions, etc. All of this can be implemented with wide-ranging effects and gravely threatens the security of the countries subjected to aggression.

Now, each new detection of a hack or blockade of internet networks indicates the vulnerability of even the most modern technologies. However, software developers have yet to demonstrate a commitment to protecting the internet. We should expect that it will be possible to conduct psychological influence against the enemy through the same channels without any witnesses and to warn our own state in advance about a threat to our national interests. Access to a global computer network makes it possible to transmit the necessary information to any region of the world and to perform many tasks associated with information confrontation.

In light of the fact that we are already seeing the movement of some countries toward contactless wars, we should expect that these countries will stake a lot on information support: of military-technical superiority; in information warfare, on control systems for strategic offensive and defensive forces of various levels; of systems of high-precision offensive and defensive weapons; the creation of a complex radio-electronic environment in the airspace in the combat area of operations and throughout the theater of war (military operations); by forcing the enemy to conduct military operations in their favor due to their information superiority and the capability to provide information support to mass high-precision missile strikes in all directions.

Information confrontation is multifaceted. Using a systemic method, one can quickly find the most vulnerable spots in the control systems, communications, computer support, reconnaissance, and all-around support of enemy combat operations. By disabling the critical backbone elements of the enemy's military and economic infrastructure, one can significantly increase the effectiveness of one's actions in other types of confrontation. The critical links of the enemy's control system primarily will be information assets, the suppression, destruction, or annihilation of which will lead to an immediate decrease in the enemy's ability to control combat systems, forces, and assets, and, therefore, to deliver mass high-precision missile strikes against targets of economic potential.

Electronic suppression will likely remain the most important component [of information confrontation]. It is already one of the most effective types of combat support in modern warfare. In contactless wars, electronic suppression will certainly shed the status of combat support and become an independent type of confrontation, especially in cases where opponents will wage wars of different generations.

The aggressor will deliver continuous mass strikes with high-precision cruise missiles and other missiles against the territory of a technically lagging enemy. [These effects] can be created by a storm or even a hurricane jamming the electronic environment. As a result, absolutely all radio-electronic assets on the ground, on the water, underwater, in the air, and in space will be blocked. Those assets that will continue to function and emit electromagnetic energy will be destroyed immediately by homing anti-radiation missiles.

Between 2030 and 2050, we should expect significant breakthroughs in the field of information confrontation. During this time, artificial intelligence—which is likely to find widespread application both in offensive and defensive weapon systems and in the forces and means of electronic warfare—can be created. Early work in this direction began in the 1960s, but at the beginning of this century, we should expect the appearance of fundamentally new electronic models of intelligence. They will probably be built like neural networks in the human brain and will be able to process all incoming information simultaneously and, most importantly, the[se models] will be able to learn. Artificial intelligence will find widespread applications; first of all, in homing warheads for high-precision intercontinental and cruise missiles and in missile defense systems.

Other efforts in this area likely will be associated with the creation of molecular computers made of organic materials combined with silicon-based circuits. The processing of information in these machines will be radically different from conventional electronic circuits. A three-dimensional protein grid is theoretically capable of processing information even faster than the human brain. Such computers are likely to be used in reconnaissance and combat systems that include land, sea, air, and space-based components.

Thus, information confrontation is practically becoming the most important element of the wars of the new, sixth, and likely subsequent generations [as well]. In the future, we should expect the use of forces and means with artificial intelligence in this confrontation. Thanks to the ability to accomplish operational and strategic tasks successfully in the wars of subsequent generations and with the help of information confrontation, information confrontation will acquire significant independence and become an integral element of all other forms of warfare.

CHAPTER THREE

Information Weapons and Information Warfare: Reality and Speculation

Today, in almost every publication and discussion, it is possible to read or overhear phrases that include the words "information" and "informational."[1] Everything has become informational: objects, processes, [and] phenomena. There is information technology and information support, information environment, sphere and space, information weapons and information conflict or war, etc. It is a real information boom. Moreover, the word "information" and its derivatives are used so often that it is sometimes difficult to grasp the meaning of what has been said, to understand where there are actual scientific problems, and where there are only superficial assessments of new phenomena; superficial generalizations, as a rule, [are] accompanied by the construction of many "new" concepts.

Such a conceptual craze can arise for various reasons—and not always objective ones. Therefore, many obviously erroneous premises can be ignored or not accorded any attention. However, more and more often, with great ease, abstract assessments of new factors are introduced in the sphere of military activity, thereby exerting a negative impact on the defense capability of the state. In particular, the absolutization of the role of information in modern military conflicts leads to a distortion of their essence and main content—armed struggle. And because the training of military cadres, future generals, and marshals—who would be able to navigate new environ-

[1] [Orlansky, 2008.]

ments as easily as A. V. Suvorov and G. K. Zhukov in their time—is dependent on this, such abstract conclusions are not so harmless. In this regard, it is important, in our opinion, without delay, to understand the situation associated with new opportunities in the field of the use of information and some new terms that can have a negative impact on the military—and in particular, operational—art. These terms include "information weapons" and "information warfare."

As is known in the Military Doctrine of the Russian Federation, the main tasks of ensuring military security during a threat situation and from the beginning of war (armed conflict) include the organization and coordinated conduct of armed, political-diplomatic, informational, economic, and other types of warfare.[2] It follows from this proposition that armed, informational, and other types of warfare are significantly distinct structural elements of a military conflict; have their own goals, essence, and substance; [and are] conducted by their own forces and means. Therefore, the statement that in the future "armed struggle will be permeated with more extensive information confrontation,"[3] seems to be inconsistent not only with the provisions of the country's Military Doctrine but also with objective reality. The dialectical interdependence of phenomena by no means presupposes their abstract interpenetration, leading to [the] deformation or complete denial of their nature, [and] the formation of some derivative forms that do not have specific characteristics.

Perhaps there is no need to specify with which forces and means armed conflict is waged in modern conditions. It merely should be noted that the effective use of any of the most powerful and accurate weapons depends on the availability of information about the position of the targets, the capabilities of its forces and assets, and other conditions, the consideration of which during conflict plays important roles in achieving success. However, although the informational characteristic of an armed confrontation always

---

[2] [President of Russia,] *Voyennaya doktrina Rossiyskoy Federatsii* [*Military Doctrine of the Russian Federation*], [Moscow,] April 21, 2006; *Armeiskii sbornik [Army Digest]*, Vol. 6, No. 7, 2000.

[3] M. A. Gareev, "O kharaktere vooruzhennoy bor'by budushchego [On the Nature of the Armed Combat of the Future]," *Vestnik Akademii Voennykh Nauk [Journal of the Academy of Military Sciences]*, No. 2, 2005, p. 13.

had an important—sometimes even decisive—influence on its outcome, this characteristic has never changed the essence of an armed conflict, [and] has not turned it into an informational [one]. This is because of the fact that information, at all times, has played a specific role not by itself, but only in connection with processes, forces, and means in whose interests it was used. In other words, information has always played a supporting role and could be of decisive importance only with all else being equal, which, in an armed conflict, includes the weapons of the opposing side and troops capable of using them.

The situation is similar today. For all the importance of information, it does not replace, and possibly will never replace, weapons, and it will not become the main means of conducting a struggle to defeat.

Obviously, therefore, the term "information weapon" is interpreted, as a rule, very broadly, and the use of such weapons is considered to be an instrument of covert economic and military-political pressure.[4] At the same time, the development of so-called unifying[5] concepts that are based on overly broad generalizations makes it possible for some researchers to say that information weapons include not only a set of tools that allow certain actions with information but also information itself.[6] This position, unfortunately, has turned out to be very resilient despite its vulnerability and obvious bias.

The conventional wisdom that a word can kill or, conversely, cure an illness with emotional intelligence, sounds very impressive. However, one cannot be guided by emotions while speaking about the effectiveness of the use of certain types of weapons when considering complex issues of theory and practice; in particular, armed conflict. With all the power of a word, which carries in itself murderous or healing information, in our opinion, today it is not necessary to speak of information as a weapon. Information has not yet become a means that can affect a person as effectively as modern weapons. It is unlikely that this [statement] must be justified. The question

---

4 A. F. Andreev and I. S. Belobragin, "Informatsionnoye protivoborstvo i bezopasnost' gosudarstva [Information Confrontation and State Security]," *Vestnik Akademii voennykh nauk [Journal of the Academy of Military Sciences]*, Vol. 4, No. 17, 2006, p. 22.

5 Andreev and Belobragin, 2006, p. 22.

6 Andreev and Belobragin, 2006, p. 24.

is whether information will ever become capable of exerting the same effective impact as other means of warfare.

The fact is that desire for world domination, in the form of an obsession, from time to time captures the imagination of the military-political leaders of some states. For this purpose, more and more effective weapons were developed and supplied to equip troops, right up to nuclear [weapons that are] capable of destroying all life on the planet. Today, new technologies open up wide opportunities for humanity, including the creation of fundamentally new means of destruction or [the creation of] such an impact that it will ensure the achievement of the most-ambitious goals in interstate or internal military conflicts. At the same time, more and more often, one can hear about the prospects of using information for [achieving these goals].

In this regard, the question of what exactly is meant [by *information*] is of fundamental importance. Today, some scholars understand information as a fundamental, generalizable, beginningless [and] endless *lawmaking process* [that involves] relations, interactions, interconversions, mutual preservation of energy, motion, mass, and anti-mass based on materialization and dematerialization in the micro and macro structures of the universe.[7,8] On the basis of such definitions, it is argued that information is the fundamental principle of the universe and the world, the primary cause, essence, source, and carrier of all phenomena and processes, all material particles and objects.[9] Probably, with a sufficiently deep justification for this definition and its recognition on a general scientific level, the question of the further development of science in general and military science in particular would be put on a different plane than today.

However, one should not forget that such an attitude toward information does not go beyond a scientific hypothesis, which cannot be sufficiently

---

[7] Ivan Iuzvyshyn, *Informatsiologiya ili zakonomernosti informatsionnykh protsessov i tekhnologiy v mikro- i makromirakh vselennoy [Informationology or Patterns of Information Processes and Technologies in Micro- and Macroworlds of the Universe]*, Moscow: Radio i sviaz, 1996.

[8] M. Poteev, *Kontseptsii sovremennogo iestestvoznaniya. Uchebnik [Concepts of Modern Natural Science. A Textbook]*, St. Petersburg: Piter, 1999.

[9] V. Tsygankov and V. Lopatin, *Psikhotronnoye oruzhiye i bezopasnost' Rossii [Psychotronic Weapons and Security of Russia]*, Moscow: Sinteg, 1999, p. 40.

substantiated. At the same time, the importance of [information confrontation] required not only clarification of the term "information" at the level of explanatory dictionaries of the Russian language, but also decisionmaking on this matter at the legislative level. That is why the basis for scientific, applied, and practical work in modern conditions is law, in which *information* is defined as information about persons, objects, facts, events, phenomena, and processes, regardless of their presentation.[10] This distinction allows us to consider information as an ideal object that does not exist without a physical medium. As information (and not a lawmaking process and not the fundamental principle of the universe), information can have a certain impact on an individual and on the masses, the impact of which is now estimated to be relatively low.

Thus, according to scholars of the Russian Academy of Military Sciences, "waging a victorious information war is still an intractable task . . . Currently, we are seeing only attempts to introduce means of guaranteed impact on the individual and mass consciousness. . . ."[11] It should be clarified that, in this case, the authors mean specifically those information wars that have been waged in society for millennia and in which, as the main influence factor—influencing the course and outcome of events—is none other than information.

Taking this into account, it is premature, in our opinion, to put on the agenda the question of how much, in the foreseeable future, the role of information will increase as the main factor of influence [and whether] it will replace the types of weapons used in modern wars. Here, all paths are open. It is possible that the direction associated with the development of the theory of *conscientious war*—wars of worldviews[12]—will turn out to be so promising that ultimately the application of methods of manipulating individual and public consciousness will become as simple as delivering missile

---

[10] Federal Law No. 24-F3, "Ob informatsii, informatizatsii i zashchite informatsii [About Information, Informatization, and the Protection of Information]," Moscow, February 20, 1995.

[11] Andreev and Belobragin, 2006, p. 24.

[12] Andreev and Belobragin, 2006, pp. 26–27.

and air strikes against troops. However, it is premature to speak of the availability of such methods.[13]

If such a hypothetical possibility ever becomes a reality, then we will no longer have to speak of any other types of weapons except for information, or any other wars except information, and prospects for the continued existence and development of society will be considered from completely different positions.

So, it is not yet possible to rely on information as a tool that can be used instead of modern weapons, but because attention to this issue is not diminishing, we will leave it for consideration by future generations of researchers.

Today's more relevant issue is that information weapons are divided into technical (cybernetic, etc.) means of information impact and "actual information means, associated with the perception and impact of information on the individual and society."[14] It is not difficult to see that, in this case, we are speaking of an information weapon that only partly correlates with traditional information wars, because, as is known, technical means of information impact were not used [in such wars]. Now, speaking of information weapons, this often means not only and not so much information as technical, cybernetic, and other means that enable impacts on information itself—"transmitted, processed, created, destroyed, perceived, and stored."[15]

This position represents information struggle on a completely different plane, opening up broad opportunities for violating logic and the formation of many new concepts that are inconsistent with the essence of the [information] phenomenon. And although the same source says that, instead of information weapons, one should speak of "cyber weapons" as more correctly reflecting the essence of what is happening,[16] its authors, unfortunately, did not consider it necessary to concretize some fundamental points in relation to the modern conceptual apparatus [e.g., a framework for thinking about information confrontation].

---

[13] Andreev and Belobragin, 2006, p. 26.

[14] Andreev and Belobragin, 2006, p. 25.

[15] Andreev and Belobragin, 2006, p. 24.

[16] Andreev and Belobragin, 2006, p. 24.

However, further dwelling on abstractions is unlikely to contribute to significant progress in an objective assessment of the concept of "information war" and its impact on society. In this regard, it is necessary to emphasize the fundamental point that information cannot directly be an object of influence because it can only be influenced indirectly through its carrier. This is due to the fact that information is an ideal object that does not exist in material form. It can be removed *from the carrier or together with the carrier*; destroyed ("erased") *on the carrier*, such as a magnetic [disk]; or destroyed *together with the carrier*.

It follows that, speaking of the impact on information in the course of information warfare, we mean, first of all, the impact on certain material objects. Therefore, depending on the established goals, technical (cybernetic and other) impact on information carriers can lead to distortion (loss) of their functions in relation to information (when the carrier becomes permanently or temporarily unable to carry out certain information processes with the required quality) or to their destruction together with information.

This is precisely how issues are resolved in the course of an armed conflict, when, while carrying out missions with troops to defeat important [targets], informatized enemy targets are subjected to the complex effect of various types of weapons. If it is required to impact only the information on those objects, then specific tasks are carried out within the framework of reconnaissance, electronic warfare, and other types of operational support. Here, it is appropriate to mention the need to draw attention to the development of methods for performing these tasks under new conditions in the interests of increasing the potential capabilities of troops. One of the reasons for the lag in this area is precisely the ambiguity in the interpretation of information warfare, its relationship to armed struggle, and the role of both in military conflict.

It is noteworthy that the use of technical and cybernetic means in information warfare to physically influence information (i.e., on its carriers) transforms this struggle into an armed one, but [one that is] waged with weapons that are different than traditional [weapons] and [the features of the struggle are] created based on the latest technologies. This is confirmed by unequivocal statements by a number of authors, indicating that, with direct physical impact, it is not entirely correct to speak of an information weapon because, in this case, an aerial bomb or artillery shell will also have to be recognized as an information weapon.[17] Naturally, such a "generalizing" concept is objectively unacceptable for defining modern weapons. And because an impact on a material object (the information carrier) by technical and cybernetic means is always a physical impact, it should be attributed to the content not of informational conflict but of armed conflict, while these means themselves are promising types of weapons. This is consistent with the proposed classification of modern types and means of destruction.[18]

It is well known that in the course of cognition and transformation of reality, it is important to take into account the continuous integration and differentiation of sciences and processes, not absolutizing each of these [processes], but finding their optimal combination for a given period of time. Ignoring this approach leads to an increase in the role of subjectivity and voluntarism in decisionmaking. This is illustrated by the excessive enthusiasm for integration processes, unreasonable generalizations, and overestimated assumptions in the field of research of new phenomena associated, in particular, with the development of digital technologies.

In a number of cases, it is quite obvious that over-generalization, with the slight distortion of the meaning of new concepts, led to difficult-to-eliminate negative consequences that could lead to deadlocks in important areas of scientific development, including [the development of] military [science]. It would be worth drawing the attention of the public and scholars to new information technologies, unambiguously defining them as *digital*, which is more in line with their essence.[19,20] Then the word "information" would not have become a household word, which turned out to be the

---

[17] Andreev and Belobragin, 2006, p. 24.

[18] A. Tasbulatov and V. Orlansky, "Razrabotka sovremennoy klassifikatsii vidov i sredstv porazheniya—neotlozhnaya zadacha voyennoy nauki [The Development of a Modern Classification of Types and Means of Destruction Is an Urgent Task of Military Science]," *Voennaya mysl' [Military Thought]*, No. 4, April 2007, p. 60.

[19] A. Akulinchev, "Problemy tsifrovizatsii voyennykh setey svyazi i puti ikh resheniya [Problems of Digitalization of Military Communication Networks and Ways to Solve Them]," *Voennaya mysl' [Military Thought]*, No. 9, September 2006.

[20] Y. Brammer and I. Pashchuk, *Tsifrovyye ustroystva [Digital Devices]*, Moscow: Publishing House "Higher School," 2004.

cause of many violations of elementary logic in the course of discussions [and] scientific research in the field of theory and practice, including military affairs.

In this regard, the very idea of mechanical, social, and other systems as information systems seems to be incorrect.[21] If the modern world is changing significantly as a result of the development of digital technologies—some on subjective grounds, called informational—then why, for this reason, change the names of traditional systems? After all, tomorrow, in connection with the development of nanotechnology, these same systems (the essence of which will not change) will need to be called nanosystems, and armed conflict, which today is being transformed into informational [systems], will need to be called nanodefense. This position seems to be very far from the objective-dialectical [one].

To some extent, it is unacceptable to assign technical means and cybernetic weapons to the field of information warfare.[22] Everything that concerns the application of these means, and even more so psychometric weapons, which includes any means intended for *violent influence* on the brain of a person or a mass of people, as well as on a variety of other material objects of the world around us,[23] cannot have anything to do with information warfare. Indeed, already at the present stage, only isolated attempts to use the latest technical means for criminal purposes—for example, in the computerized financial sphere—create a resonance all over the world [and] provoke sharp and powerful responses from the global community, politicians, statesmen, and business circles.

The use of tools [that are] developed using the latest technologies for aggressive purposes is outside the framework of international law.[24] And because, in their potential capabilities, [these tools] will be commensurate with weapons of great strength, sooner or later it will be necessary to

---

[21] Andreev and Belobragin, 2006, p. 23.

[22] Andreev and Belobragin, 2006, p. 25.

[23] Tsygankov and Lopatin, 1999, p. 17.

[24] V. Akhmadullin, "Kiberprostranstvo pod pritselom Pentagona [Cyberspace at Pentagon's Gunpoint]," *Nezavisimoe voennoe obozrenie [Independent Military Review]*, No. 1, January 12, 2007.

establish a legal basis that limits or completely prohibits their use, at least in peacetime. This is because of the fact that, when such weapons are created on a mass scale, a clash of armed forces equipped with them or other power structures of opposing states will be nothing more than a military conflict with the use of fundamentally new weapons. It is probable that such conflicts, depending on their scale, will differ significantly from modern ones in character, content, and consequences.

The foregoing allows us to reach the following conclusion: Informational [influence] only can be considered impact through information, but not impact on information. Influence with the help of information can only be exercised on a person because it can be perceived in the form of information only by a person (we will not take into account the animal world). It follows from this that information warfare (according to the essence of the given phenomenon) cannot include any technical (power) aspect. In this conflict, information is used as a factor of influence on individual and public consciousness and, at the same time, as a means of protection against such influence.[25]

The means of conducting such a struggle are, first of all, mass media, which provides information from each of the opposing sides to the specific targets. In this case, information superiority can be interpreted as a more effective influence with the help of information. To gain such superiority requires better information and better ways of communicating it, connected to the development of global communication systems. Already today, to conduct information warfare, some states widely use the internet. In particular, "American specialists in the field of informational-psychological influence already have experience working in the computer networks of the adversary."[26]

A different (forceful) approach to the interpretation of the concept of "information warfare" leads to intractable contradictions between views of informational and armed conflict, which is the subject of military—and in particular operational—art. The inclusion of a force aspect in information warfare would inevitably lead to the distortion of established practices

[25] Tsygankov and Lopatin, 1999, p. 14.

[26] Akhmadullin, 2007.

FIGURE 3.1
General Structure of Information Warfare

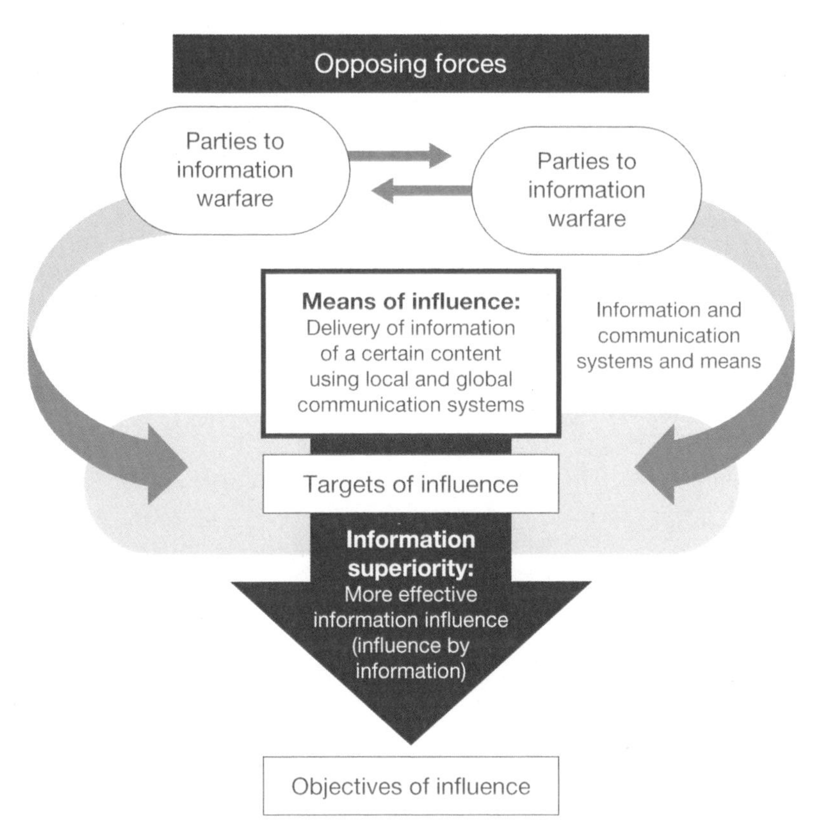

of the theory and practice of operational art, which could have a negative impact on the training of military personnel and on their mastery of the already complex modern forms and content of armed conflict.

Summarizing the previous, the following conclusions can be drawn.

First: Information weapons and information warfare today are conditional categories because information has not yet become a means that is comparable, in terms of the strength of [its] impact, with traditional weapons, and information warfare is not understood as a confrontation in which information is an influential factor.

Second: In modern conditions, the widespread use of the media, including the internet, for solving information warfare tasks, is a factor in the impact of information. This requires the development of views on the content of such a struggle and its conduct using the appropriate conceptual apparatus.

Third: Information warfare—conducted solely using information for the purpose of informational-psychological influence—should be separated from psychological warfare, for which there is a variety of forceful methods and means that are not related to information. The psychological struggle waged using these means, in contrast to the informational [struggle], is directly related to armed struggle.

Fourth: The inclusion of the technical or cybernetic aspect in the content of information warfare is theoretically erroneous because, in this case, the content of the war will not correspond to its name. At the same time, the line between the concepts of "information warfare" and "armed struggle" is easily erased, and there arise artificially created contradictions in views on the theoretical foundations of both types of warfare.

Fifth: The absence in theory of clear lines between different types of conflict cannot be compensated for in the course of the educational process by any pedagogical means. In the process of training military personnel, extremely clear [and] specific guidelines are required. Therefore, it is very important to develop military science and construct the educational process exclusively on objective foundations, minimizing the subjective (in the negative sense of this word) factor.

CHAPTER FOUR

Peculiarities of the Modern Period of Informational-Psychological Confrontation

Global Information Confrontation of a New Type

In today's environment, continuing technological progress contributes to the continual increase in the volume and speed of information dissemination.[1] The capabilities of information coverage of large territories and populations within the shortest possible time frame are being enhanced. Along with the positive effects of global informatization, the contours of new international problems emerge more clearly. First of all, [informatization] applies to the domains of information security and information confrontation. It should not be claimed that these problems emerged only with global informatization, but that the rise of a unified information space made it possible to turn [the information space] into one more field for confrontation in international relations.

It is becoming increasingly evident that government is dependent on capabilities to carry out information confrontation, both in domestic and foreign policy domains. The main trends of changes in the nature of geopolitical tensions between countries [and] the development of the process of globalization in the beginning of the 21st century indicate that, along with conventional methods of force and means of solving problems in [the geopolitical sphere], information is increasingly used.

---

1 [Prysiazhniuk, 2012.]

The national and transnational media, along with any other information networks [that are] capable of influencing the worldview, political views, legal consciousness, mentality, spiritual ideals, and values of the individual and society as a whole are the main instruments of information confrontation.

Under the influence of a set of objective and subjective factors, the theory of information warfare has undergone a complex evolutionary path: from the perception of it as an aid in accomplishing the tasks of combat at a tactical level to giving it a global function of managing political processes at a strategic level.

A set of measures—which, in the 20th century, became known as special information operations—[has] accompanied military actions since prehistoric times. For example, body painting for combat by ancient soldiers served both to raise morale and to disorient the enemy regarding the composition of the warring parties. It was supposed to lead the adversary to believe that it was not an army but rather supernatural, mystical forces. It is known that the purposeful disinformation of an enemy about the composition, location, and size of enemy troops is still one of the most important elements of combat.

In the 4th century B.C., the first fundamental work of the brilliant Chinese strategist and thinker Sun Tzu, *The Art of War*, appeared, in which he wrote:

> The means by which educated rulers and wise commanders acted and subdued others, and their achievements took precedence over the many, was antecedent knowledge. Prior knowledge cannot be obtained from demons and spirits, from phenomena or celestial signs; it should be obtained from people who know the true state of the enemy.

Sun Tzu first suggested the use of informational means as an alternative to armed combat. He formulated the nine commandments; following [them] would allow such a powerful influence on the spiritual world of the enemy's army that it would simply "decompose," even prior to battle. In essence, the doctrine of Sun Tzu forms a foundation for contemporary acts of informational influence and [special information operations]. But before returning to the principles of this ancient Chinese strategist and thinker, the technologies of information influence underwent a long path of development.

Initially, acts of information influence were treated as a means of military-political disorientation of the adversary and were contained to two main approaches:

1. disinformation about one's own resources
2. actions aimed at failure or obstruction of data transmission channels to disorient and disorganize the adversary (as a result of successful implementation of such operations, the enemy loses the ability to act in a coordinated manner, which significantly increases the vulnerability).

To perform the abovementioned tasks (particularly the second approach), the armed forces of the countries include radio-electronic warfare troops. [These forces] engage in the creation of noise over radio channels, interfering with the functioning of adversaries' electronic reconnaissance means. In offensive operations, the targets to be destroyed as a matter of priority are communication nodes—telephone switch stations, servers of cellular operators and internet providers, and television and radio transmission stations. In most cases, the attack is preceded by the identification of all sources of electromagnetic fields that could be used as means of communication.

The set of actions described previously belongs to the so-called first-generation information weapons, or to the technological [technogenic] side of information confrontation. Over time, in parallel with the development of [the] technological [technogenic] [side], the human side [of information confrontation] also started to take shape. Increasing attention is paid to the need to influence the psychological traits of the leadership and personnel of enemy troops and the civilian population in the rear.

Beginning in the 1950s, Pentagon experts started to consider aviation flyovers as psychological weapons. During the Korea campaign, a concept was developed that significantly increased the effectiveness of bombings because of their (seemingly collateral) psychological effect. The chaos and panic caused by the bombings came to the forefront of the conventional tasks of destruction of manpower, military, industrial, or civilian objects. Bombing came to be seen not as a means of destroying the enemy's infrastructure before a ground invasion, but as a mechanism with which it was possible to persuade the adversary to cease [its] resistance.

Undoubtedly, the will of the enemy to resist can be broken not only by missile attacks. A relatively new direction, termed *propaganda*, was already developing rapidly in Europe in the 1930s, [when] the first, relatively successful instances of using the media as a weapon were carried out. By the start of World War II, the Germans were the absolute leaders in the field of information technologies. German specialists managed to organize the internal information space in such a way that external infogenic [information] threats simply vanished within it. Citizens of the Third Reich were convinced of the invincibility of their country and the sacredness of the values on which it was based. An entire industry was created that was responsible for the content of the information field: print media, radio, cinema, recording studios, visual advertisement, and so on. The Third Reich lent serious effort to the creation of a new culture and the means for its dissemination. German shortwave radio purposefully worked with foreign audiences, instilling in them the idea of Germany as the most progressive and invincible country. It should be noted that the further development of similar systems in other states was a reaction to the activity of the Germans. German propaganda technologies were adopted by such countries as the [Soviet Union], the [United States], and Great Britain, etc. At the beginning of the Cold War, the media—and especially radio as a means of informational-psychological influence—was used by almost all states that were involved in confrontation (both leaders and satellites).

Thus, at the beginning of the last quarter of the 20th century, informational-psychological confrontation took shape as a set of technogenic [technological] and humanitarian levels that helped perform the following tasks:

- blocking or damaging the adversary's command and control channels
- misinforming the enemy
- creating an atmosphere of tension and panic from the constant fear of attacks in the enemy's rear
- influencing the social consciousness of the enemy with the aim of demoralizing the adversary.

In the period described, the classical approach to special information operations was based on purely linear views, according to which the result of external command is unambiguous and linear, [and] propagandistic efforts work according to the scheme: Control through influence [is] the desired outcome. Thus, an increase in the intensity of the influence always increases the return. Nazi minister of propaganda J. Goebbels called it a "harsh interpretation." During the 20th century, the principle of "harsh interpretation" was considered a universal paradigm of information policy. However, since the 1980s, the West, for the first time, sensed that the effect of influence was not always directly proportional to its intensity.

The crisis of the classical concept of propaganda coincided with a new wave of informatization, or, to be more precise, was the result of it. A linear approach to the execution of information operations, which assumes a direct link between the intensity of the influence and its outcome, was less and less relevant under the new circumstances. [The] development of information technologies led to a significant increase in the number of available information channels. The ability to have a dialogue that was revived in communicative processes formed a natural barrier to the perception of propaganda. Even the most powerful propagandistic impulse simply faded in a system that was expanding and branching out at an accelerated rate. The increase in the number of channels involved the formation of so-called subrealities. The recipient could now choose the source of the interpretation of reality, which would correspond to [the recipient's] own worldview. It can be stated that the information system began to take shape using the social system as an example—as a key carrier of social consciousness, or, in other words, a kind of virtual projection of society. Naturally, under such conditions, a symmetrical, linear approach both to the study of information and communication and to the organization of information operations had to be replaced by something more appropriate.

Research on changes in the nature of the relationship between the media system and consumers of information underwent significant transformations: An attempt was made to explain [changes in the relationship] from the perspective of synergetic laws, which later became part of the arsenal of mass communications specialists as a basis for modern information policy. Society came to be seen as an extremely complex system, each of elements of which had many degrees of freedom, [and] in other words, constituted a system that was in a state of constant hesitation before choosing one of the possible evolutionary paths. The choice of a developmental path, according to modern ideas, can be affected by an impulse of even minimal tension.

On one hand, new conditions significantly complicated the tasks of information operation specialists, but on the other hand, these conditions brought the potential capabilities of "information warfare" to a fundamentally new level. If before information operations were described only as accompanying combat and having an ancillary function to conventional military force, under the new circumstances, an information strategy can replace conventional methods.

It is believed that Ukraine, in its current state, has cost the United States $14 million [in terms of military and security assistance]. However, the amount spent in Iraq, as of 2007, exceeded $450 billion. The words of U.S. President [Richard] Nixon, proclaimed in the 1960s, have turned out to be prophetic. He said that he believed that one dollar invested in information is more valuable than ten dollars invested in the development of weapon systems, as the latter are unlikely to ever be applied, while information works continuously and everywhere. Military invasion is gradually becoming a last resort, used only when information operation specialists cannot fulfill the tasks.

Recently, as part of the "Information Revolution" initiative of the Strategic Estimates Program of the U.S. National Intelligence Council, an analytical corporation, RAND (RAND Corporation) conducted a number of international scientific conferences and seminars, during which the opinions of prominent experts on the problem of the transformation of society under the influence of the information revolution were studied and assessed.

The results of the work performed were summarized by RAND experts in a report, *The Global Course of the Information Revolution: Recurring Themes and Regional Variations* (MR-1680-NIC), [which was] published in summer 2003.

The main purpose of the study was to identify the nature of the impact of information technologies and the information revolution on economic, financial, political, cultural, social, and other domains of modern society, and to forecast the situation for the next ten to 20 years.

The research noted that [as of the report's writing], the progress in information technologies ha[d] already affected most domains of business, government, and public affairs in almost all regions of the world. Information technologies and a concomitant information revolution turned into one of the most significant factors that contribute to the dynamic transformation of society [and] its transition from post-industrial to an information [society].

The research results made it possible to identify the peculiarities of the development of information technologies and the impact of the information revolution, including both those [technologies] that are relevant for most regions of the world, and specific ones [that are] pertinent to certain regions of the planet.

Thus, for most regions of the world that seek to take advantage of information revolution accomplishments, the RAND experts identified the following characteristics [of the development of information technologies]:

1. The development of new technologies will continuously spur the information revolution.
2. The information revolution will generate new business models that will significantly transform the business and financial world.
3. The information revolution will significantly affect mechanisms for managing society and will create new political players.
4. The information revolution will remain multifaceted and will be shaped by social and cultural values.
5. The multifactorial form and feature of the national approach toward the perception of the information revolution will be preserved.

In addition, RAND experts predicted the following main trends in the development of the global geopolitical situation:

1. For the next ten to 20 years, the United States will remain at the forefront of the information revolution.
2. The information revolution in Europe will develop more slowly and in a different way than in the United States and Canada.
3. In the next ten to 20 years, several countries in the Asia-Pacific region will continue the rapid development and large-scale use of information technologies.

4. Geopolitical tendencies [that are] supported by the information revolution may present new challenges to the national security of the United States and other developed countries of the world.

Because the pace of these technological revolutions and their synergetic impact is growing, an understanding of the consequences of their influence on future society is also increasing. RAND experts stated that, during these technological revolutions, the inequality of individual nations and regions of the planet will be preserved; moreover, the acceleration of the technological revolution will lead to deepening inequality and, as a consequence, to an unprecedented increase in tension all over the world.

Today, according to American experts, information confrontation is not simply a means of supporting the operations of the armed forces by disrupting processes of command and control of troops, but goes far beyond these issues. This is evidenced by the main results of studies [that were] conducted by American experts from the RAND Corporation in the late 1990s. In this study, the term "strategic information warfare [confrontation]" was used for the first time. This kind of confrontation, according to the authors of the report, is [defined as] the usage by the countries of the global information space and infrastructure for strategic military operations and reducing the influence on one's own information resource[s]. This research allowed [us] to distinguish the main features of this type of confrontation: the relatively low cost of creating means of information confrontation and the collapse of the status of traditional state borders in the preparation and conduct of information operations.

Further research on the problem led to the introduction of the concept of "second-generation strategic information warfare [confrontation]." In the report, it is defined as a fundamentally new type of strategic confrontation, spawned by the information revolution, which introduces the information space into a variety of possible domains of confrontation. It is emphasized that the development and improvement of approaches to managing second-generation strategic information confrontation may lead to total renunciation of the use of military force in the future. Essentially, the second-generation information confrontation is reduced to efforts of transforming the adversary, to destroy its traditional meaning and fill it with the new one. That is, it is not about instilling certain ideas in individuals or groups, but about the development of a full-fledged social worldview that has the ability to self-develop in the right direction.

From the early 1970s to the late 1990s, the Americans were the absolute leaders in the field of information confrontation. They seemed to successfully fulfill the task of intensifying the use of global information resources and blocking their resources for other countries. As a result, in 2003, according to Chinese experts, the flow of information from developed to developing countries accounted for 80 percent of all informational exchange between them.

Currently, the United States is intensifying the work aimed at implementing national information strategy to ensure information advantage by imposing information that would encourage the higher military-political leadership of other countries to make decisions [that are] favorable for the United States. [The] main elements in achieving the goals of the national information strategy are managing the perceptions of the target audience and shaping "public opinion" by manipulating information.

U.S. national policy goals will be achieved through strategic information confrontation with the use of offensive information weapons. Recently, [it is] not hardware/software means of influencing information systems and information resources of the adversary but the means and methods of manipulating information that are being increasingly considered. This is evidenced by [the] analysis of developments on this topic—the number of publications in the foreign press on the development of means and methods of manipulating the consciousness (in particular, neuro-linguistic programming, hypnosis, and other suggestive methods), research on personality psychology, etc., [that] has recently increased. A number of new notions have emerged, such as "real virtuality," when the coverage of a certain event in the press becomes more important than this event itself.

Scholarly research on such programs as "MK-Ultra," "ARTICHOKE," etc., that were conducted in the 1960s and 1970s established that the most-promising methods of information warfare are the methods of influence on individual, group, and social consciousness. Implementation of such methods at the state level requires review of the main approaches to conducting foreign and domestic policy in the information age.

The main forces involved in contemporary strategic information confrontation will be small groups of highly qualified political technologists [specialists in political manipulation], speechwriters, and image consultants who create and produce given scenarios. Today, the work of such experts is called "public relations." A group of such experts, led by Jamie Shea, covered the conflict in Yugoslavia. It was in Yugoslavia where the entire cycle of strategic information confrontation measures was fully tested: from discrediting the political leadership prior to the conflict to [providing] favorable coverage of the events of the armed aggression.

It should be noted that, in many respects, the processes of globalization are objective and caused by the level of scientific and technological progress; abandoning advanced achievements today is simply impossible. The United States and other developed countries were among the first to realize the benefits of globalization and [try] to build a model of new global society in accordance with their own, largely selfish, interests. However, the vulnerability of this idea is evident: A stable global society can be built only on the basis of a network—not hierarchical—structure, where each node would be equal [with others]. Under these conditions, one of the pressing problems is the development of new ideas for the further positive development of global society.

Information Weapons Under Modern Conditions

The third millennium is marked by the rapid global development of computer information technologies, means of electronic telecommunication, and their introduction into all spheres of social activity. The rapid pace of development of the computerization and informatization of society inevitably leads to the creation of a unified world information space, in which all means of collecting, accumulating, processing, exchanging, and storing information are accrued.

The information space, in fact, becomes a theater for military actions, where each opposing side seeks to gain an advantage and, if necessary, defeat the adversary. Confrontation in the information domain reached such a scale that it required the creation of a special notion called "information warfare" or "information confrontation."

The first works on the notion of "information warfare" began in the United States in the early 1990s. Currently, there are several interpretations of the term "information warfare." The differences between them are insignificant, so one has a good reason to use the version presented in the U.S. Army Field Manual 100-6, *Information Operations* (published in August 1996).[2] According to this document,

> information warfare is a set of actions taken to achieve information superiority by affecting information, informational processes, information systems, and computer-based networks of the adversary while defending one's own information, information-based processes, information systems, and computer-based networks.

Actions within information warfare can be both offensive and defensive. Accordingly, the existing defensive and offensive means of information confrontation are being enhanced and new ones are being actively developed to achieve information superiority over the enemy. An information weapon is a means of information warfare or confrontation.

To determine whether the notion of "information weapon" should exist, it is necessary to first address the definition of a weapon. In the *Soviet Military Encyclopedia*, a weapon is construed as "devices and means used in armed combat to defeat and destroy the enemy."[3] The key part of the definition of weapon is the purpose of its use—to defeat the enemy. Objects (targets) are defeated if they are being acted on by various means and, as a result of these actions, they completely or partially (temporarily) lose the ability to function normally (execute a combat mission). Defeat of objects implies their annihilation (destruction), suppression, and depletion (of the manpower of these objects).

Destruction of an object implies inflicting such damage on it that it completely loses combat effectiveness.

Suppression means the damaging (harming) of the object and creating such conditions under which [the object] is temporarily deprived of combat capability, its maneuver is limited (forbidden), or its control system is broken.

---

[2] [U.S. Army Field Manual 100-6, *Information Operations*, Washington, D.C.: Headquarters, Department of the Army, August 27, 1996.]

[3] [*Soviet Military Encyclopedia*, Moscow: Voenizdat, date of publication not supplied by original author.]

Depletion is the prolonged firing at an enemy with a limited number of forces or means of conducting periodic air strikes. Its main purpose is to morally and psychologically influence the manpower of the [target] and thereby reduce its combat effectiveness and normal functioning.

So, is the so-called information weapon capable of striking the enemy?

Information weapons, according to one of the existing definitions, are a set of software and technical means designed to control information resources of the object of influence and interfere in the work of their information systems.

Information weapons can be classified according to methods of influencing information, information processes, and enemy information systems. This influence may be *physical, informational, software-technical,* or *radio-electronic.*

Physical influence can be exercised by applying any methods of defeat by fires attacks. However, it would be more correct to include within information weapons of physical impact the means designed solely to affect the elements of the information system: anti-radar missiles, specialized rechargeable batteries for generating high-voltage impulses, means for generating electromagnetic impulse, graphite bombs, [and] biological and chemical agents [that influence] the element base.

With the help of anti-radiation missiles in the first days of the air operation of the coalition of peacekeeping forces in the Persian Gulf (1991), 80 percent of Iraq's ground-based radars were disabled.

The use of electromagnetic radiation generators is also effective. Experiments have shown that simple, small-sized generators used at a distance of up to 500 meters can cause dangerous damage to an airplane's controls during take-off or landing and can shut down the engines of modern cars [that are] equipped with microprocessor technology.

Graphite bombs were used by the U.S. military during the Gulf War and in [the military operation in] Kosovo. [These bombs'] striking effect is achieved by the creation of clouds over an object with an area of up to 200 square meters with the help of thin superconducting fibers made from carbon. When fibers collide with the current-carrying elements (insulators, wires, etc.), the power grids are short-circuited and fail.

Biological agents are special types of microbes that can destroy electronic circuits and insulation materials used in [radio]electronics.

Information methods of influence are implemented using all media and global information networks, such as the internet [and] voice disinformation stations.

Because people, motivated by their physiological, social, and informational needs, comprise the main element of the information infrastructure, the accurate calculation of the use of so-called informational-psychological methods of influence has a direct impact on the level of security of the state. Scientific and technological progress in the field of information technology [and] the development of the media has erased national borders in the information space and created unprecedented opportunities for suppressing the enemy through unconventional means of defeat that do not cause physical destruction. Penetrating the consciousness of each member of a society [and] prolonged massive informational-psychological influence of a destructive nature pose a real threat to the existence of a nation as a result of the transformation of its historically established culture, worldview, and ideological attitudes.

The fact is that the mass media, under the guise of slogans of "objectivity of information coverage" of certain events, damages the information security of a country by manipulating information, disseminating disinformation, [and] providing information support to certain extremist and criminal groups. And this problem may be exacerbated by the monopolization of domestic media and the uncontrolled expansion of the foreign media segment within the information space of the country.

The stations of *voice disinformation* that are currently being developed in the United States will allow entering the radio network of the target of influence and generating a computer-simulated voice of the commander of the unit (subdivision) of the enemy to give orders and commands to subordinate troops, thereby interfering with their control.

The means of implementing *software and hardware methods* include computer viruses, logic bombs, and hardware Trojans, as well as special means of penetration into information networks. These tools are used to collect, modify, and destroy information [that is] stored in databases, and to disrupt or slow down the performance of various functions of information and computer systems.

Software and hardware means can be classified according to the tasks that could be performed with their help as *means of collecting information, means of distortion and destruction of information, and means of influencing the functioning of the information systems.* Moreover, some means can be universal and used both for distortion (destruction) of information and for influence on the performance of information systems of the target of influence.

Means of collecting information allow carrying out unauthorized access to computer systems [and] detecting access codes, cipher keys, or other information about the encrypted data and transmitting collected information to the interested organizations.

Currently, special software products, so-called knowbots (Knowbot – Knowledge Robot)—which are able to move through the information network from computer to computer and reproduce themselves by making copies—have been developed. A "knowbot" infiltrates computer systems and, having found information that it was interested in, leaves in this place a copy of itself, which collects and transmits this information for some time. To prevent detection, the "knowbot" may have built-in functions of self-movement and self-destruction.

The tasks of collecting information are performed with the help of software products "Demon," "Sniffers," [and] "Trap Door." The "Demon" software infiltrates into systems, records all executed commands and, at some point, transmits information about these commands. The same applies to "Sniffers" that read and transmit the first 128 bits of information that are needed to log in to the system. Programs are used to expose access codes and ciphers. "Trap Door" allows for carrying out unauthorized access to information databases, bypassing the security code. Moreover, the [computer] system and [its] security features do not identify them.

Special technical devices have been created and are constantly being modernized to read information from computer monitors. The creation of miniature, specialized systems for collecting, processing, and transmitting information that can be implemented into various electronic devices under the guise of conventional chips is also promising.

Means of distortion and destruction of information include the software products "Trojan Horse" [and] "Worm" as well as numerous computer viruses, the number of which exceeds 60,000.

The "Trojan Horse" provides hidden unauthorized access to information arrays. It is activated by a command and used for changes or destruction of information and for slowing down the performance of various functions of a system.

A "Worm" is a third-party file generated inside the information database of a system. It is capable of changing the working files, reducing memory resources, and moving and editing certain information.

The means of influencing the functioning of information systems include "logic bombs," "email bombs," etc.

A logic bomb is a set of instructions that is inactive until it receives a command to perform certain actions to alter or destroy data and create a malfunction of information and computer systems. During the Gulf War, Iraq could not use its air defense systems—purchased in France—against the multinational forces because their software contained logic bombs activated after the start of the war.

Email bombs are unauthorized messages of a large quantity that are used to increase the load on a server to make it became inaccessible or its resources insufficient for normal operations. In exactly this way, the NATO server was blocked for three days in March 1999. An unknown addressee regularly sent about 2,000 emails a day to the North Atlantic bloc's address and filled the email box.

Radio-electronic methods of influence involve the use of radio-electronic suppression, electronic intelligence, and some other [methods]. The main purpose of such weapons is to control the information resources of a potential adversary and [conduct] covert or overt interference in the operation of its control and communication systems with the aim of disorganizing, disrupting the normal functioning of, or causing malfunctioning both in peacetime and in wartime, independently or in combination with other means of influencing the enemy.

As for the mass media, using them for the purpose of an active informational-psychological influence can reduce or even deprive enemy personnel of combat capability for a certain period of time, forcing [the enemy] to evade participation in combat in various ways. In this case, the media acts as means of suppression, i.e., can be classified as the weapons.

Software, hardware, and electronic means of collecting information do not fall under the classical definition of weapons because they are not used for direct defeat of the enemy but only provide the conditions for effective use of armed [and], in particular, informational, confrontation. But taking as a basis the definition of information weapons formulated previously, the means of collecting information undoubtably provide control over the information resources of an enemy and can be included in this type of weapon.

The main ways and methods of using information weapons can include:

- damage to the physical elements of the information infrastructure (destruction of power grids, interference, use of specialized software that stimulates the failure of hardware, and biological and chemical means of destruction of the element base)
- destruction or damage of information, software, and technical resources of the adversary, disabling the defense systems, [and] infiltrating [with] viruses, hardware Trojans, and logic bombs
- impact on software and information databases and control systems for the purpose of their distortion or modification
- threat of or carrying out terrorist acts in the information space (identification and threat of disclosure or disclosure of confidential information on the elements of national information infrastructure, socially significant and military encryption codes, principles of encryption systems functioning, successful experience in conducting information terrorism, etc.)
- capture of media channels to spread disinformation [and] rumors, demonstrate power, and communicate one's own demands
- destruction and suppression of communication lines, [and the] artificial overloading of switching nodes
- influence on information and telecommunication system operators using multimedia and software tools to instill information into subconsciousness or cause the deterioration of health of an individual
- impact on computer devices of military equipment and weapons to disable them.

Thus, the development of a unified global information space as a natural outcome of the progress of world scientific and technical thought and the perfection of computer and information technologies form the preconditions for the creation and use of information weapons. The possession of effective information weapons and means of defense against them becomes one of the main conditions for ensuring the national security of countries in the 21st century.

Current processes of globalization have changed the substance and forms of information warfare. Globalization has a double influence on the nature of contemporary conflicts and wars: First, it caused the erosion of government power and [increased] social vulnerability, [and] second, it opened new opportunities and economic incentives that arise during civil war.

With this in mind, information warfare can be defined as a set of measures of information support, information counteraction, and information defense, which are carried out according to a single idea and plan with the purpose of achieving and maintaining an information advantage over the enemy. The spread of information warfare is explained by the possibility of ensuring the achievement of political goals through global (strategic) psychological operations to form an appropriate beneficial system of views [and the] psychological cultivation of the population of a country and neighboring states.

A classic example of the use of information weapons [is as follows]. During World War II, Japan took a set of measures to form the cult of "kamikaze" among military employees and the entire population of Japan. Not having any military superiority over the Americans, and delaying the inevitable defeat, the Japanese tried to intimidate the enemy with suicide attacks. As a result, the Japanese authorities succeeded in the psychological struggle—[and] retained their status in society.

The spread of information warfare is explained by the impossibility, under current circumstances, of frontal aggressive assault [and] the use of weapons of mass destruction. Therefore, information warfare allows for achieving political goals through global (strategic) psychological operations to form a positive attitude of the international community toward such actions because of the psychological cultivation of the people of the conflict areas, specifically the military and the population of the adversary and neighboring states. One's own troops are also psychologically cultivated to raise morale and form their image as liberators, bearers of democratic values, etc.

At the beginning of the 21st century, the greatest importance was placed on the image component, which implied a negative influence on the reputation of the adversary that was supposed to subsequently lead to it being ignored and discredited by the public. The last decade has seen a phenomenal increase in the capabilities of information technologies. But only now does this issue begin to appear as one of the main [issues] in the fight for the information space of the world. Information technologies could not leave unaffected such a domain of international relations as information warfare, creating a new level of information wars.

This direction was used by the United States during Georgia's war against Russia in August 2008 and during the information confrontation of Russia and Ukraine during Ukraine's "delay" of gas transit to Europe in December 2008.

Information and image wars distort reality in the social consciousness of the masses, and their outcomes may differ significantly from the outcomes of armed conflict; moreover, it can prove to be more significant than [armed conflict]. One of the postulates of behavioral sociology is appropriate here: "If the situation is determined as real, it is real by its consequences."

Information—or intangible—victory has quite tangible material outcomes. As a result of the war in South Ossetia, Russia felt the outflow of foreign capital, [and] the threat of putting American missile defense in Poland was rapidly exacerbated. Regarding the consequences of the anti-Ukrainian "gas" information campaign, it should be noted that, in addition to signing a "new" gas contract, which is difficult to recognize as accommodating the national interest of our [Ukrainian] state, the country experienced serious complications in political and diplomatic relations with European countries.

Contemporary information warfare manifests itself through the biased coverage of certain events; the widespread use of disinformation; information blackmail using the results of electronic control over people's lives, political activities, and personal plans; [and] the use of the full potential of modern media to obtain unilateral benefits.

Considering the trends of the global information society, it is impossible not to mention a phenomenon that changes the existing system of contemporary international relations. This [phenomenon] is international cyberterrorism, which, with the help of modern telecommunications, produces a "terrorist consciousness," providing an opportunity for terrorist groups to manipulate mass consciousness through the media. Media-informational terrorism became a type of informational terrorism. Through the internet, the [terrorist group] propagates its ideas on a global scale. Research shows that the role of the global terrorist environment is growing. Furthermore, internet terrorism is highly dynamic: [Web]sites emerge and disappear quickly, change their titles [and] domain names, but keep the content of links and articles on the pages of cyber publications. [Terrorists'] purposes are to influence thoughts, behavior, and consciousness or sow fear [and] panic [and] demoralize society; evoke feelings of guilt for the actions of one's own government; cause civil unrest; [and] initiate discussions of terrorism.

Most developed countries have strong information potential, which, under certain conditions, will ensure any of them the achievement of their political goals. In addition, today there are no international legal norms for informational confrontation. Monitoring of media publications convincingly demonstrates that the main trend of information confrontation is to increase its role in solving foreign policy problems. The improvement of nonconventional means at the present stage of the scientific and technological revolution led to the emergence of weapons of global destruction, the systemic application of which can destroy the habitat of mankind.

The use of information tools and systems increases the capabilities of state influence. At the same time, the vulnerability of management systems to directed influence in the information sphere is increasing. These trends objectively lead to expansion of the arsenal of methods and means of information confrontation, strengthening [information confrontation's] influence on the course and outcomes of military actions, increasing the number of used forces and means.

At the present stage of historical development, the tendency to resolve foreign policy conflicts without armed violence dominates. Information war [has] ceased to be a secondary factor, a supplement to the "main" events. It has become one of the most important mechanisms of warfare, which is being spoken about on par with the use of armed forces and equipment. In the modern world, information war has become a legitimate means of political struggle.

Despite the fact that a large part of society is aware of the process of targeted information attacks on the adversary and accepts the probability of "dirty" technologies, [information] can still be subject to manipulation by the media. As a result, it is not the speaker of truth who wins the communication conflict but the one who manages to show the audience more-exciting "information series" to justify [their] own position very clearly. That is, the larger informational capability a country has, the more likely it is to achieve strategic advantages in the future system of international relations.

Information wars have become an axiom of modern international relations and allow for achieving desired objectives quite efficiently, with few financial and human resources: It depends on the degree of professionalism of information operations implementors. Countries with a harmoniously developed and thus protected information society will find it much easier to keep their statuses.

Problems arising from the transition to an information society further exacerbate the need to understand the patterns, peculiarities, and implications of the development and employment of new media and communication means. Given the novelty, complexity, and uniqueness of the matter, there is still not enough differentiated research on the essence of information confrontation [or] methodological grounds for studying the developments in this field.

Therefore, the issue of promoting and consolidating national interests abroad has considerable scientific and applied value; primarily with regard to the research of public administration mechanisms and the development of scientifically substantiated strategy and tactics to ensure national information security.

Features of Information Security in the Age of Globalization

The change in the worldview at the turn of the third millennium was caused by a revolution in communications and the information domain. Mass computerization, implementation, and development of innovative information technologies led to impressive breakthroughs in education, business, industrial production, and research.

Until recently, both in the theory and practice of national security, the main focus was on the military component [of information confrontation]. Currently, the limitation of such an approach has become evident because the scientific and technological revolution led to the development of an information society, in which information is the main management tool and the main instrument of power.

Global social change [and] world events in the late 20th century require objective analysis of the global information environment. Before this, the issue of information security in our country [Ukraine] was not only overlooked but actually ignored. At the same time, it was believed that this problem could be dealt with by introducing total secrecy [and] various restrictions. And only in the latter years has the importance of this issue become apparent.

The current geopolitical situation requires fundamentally different approaches to the problems of national security, analysis of the content, and evolution of the entire spectrum of geopolitical factors, with information being the most important of them. Today, one can rightly claim that, all else being equal, the achievement of strategic advantage by a state depends on its information capabilities. This is proved by the results of the Cold War with the United States, which was conducted primarily by informational means: Having reached parity in the military domain, the [Soviet Union] was defeated in information confrontation. In this context, it explains the assessment by the American military-political leadership of the role of information and the reasons for the regular increase in allocations for the development and improvement of information technologies. If, in 1980, about $8 billion was spent on acquisition of information technologies in the United States, in 1994, [spending was] more than $25 billion.

With the start of the [President Ronald] Reagan administration, the theory and practice of information influence experienced radical change. The era of global competition for the social consciousness has begun with the use of the latest information technologies on the basis of coordination of activities of all state bodies and transnational corporations. The coordinated activities of informational and psychological agencies (state, public, and commercial organizations) brought about results: Nowadays, the United States dominates the global information space. And with the help of the internet, [the United States] seek[s] to establish strategic leadership in the global information space in the 21st century.

The revision of priorities and foci in the interpretation of national security matters directed the science and practice toward the need to develop an absolutely new aspect—psychological security.

Many new means of influencing people's psyches and guiding their behavior have been created. The press periodically publishes information about the American programs "MK-Ultra" [and] "Artichoke," and similar developments by France, Germany, Israel, Japan, etc.

In recent decades, capabilities for influencing the human psyche have increased dramatically. One of the main reasons for this [increase] is significant achievements [that have been made] in the fields of psychotronics, parapsychology, bioenergy, [and] other psychophysiological phenomena.

A search for new forms and methods of influencing the psyche of an individual or large mass of people is being conducted in a majority of the leading countries of the world. The United States leads on this issue, having at its disposal the most extensive network of institutions, centers, laboratories, [and] enterprises for relevant theoretical research and solving applied military problems. U.S. military agencies show great interest in these developments.

It has become technically possible to influence the human psyche via satellites. In this regard, particular concern is caused by the deployment of the satellite system "Teledesic," which would be carried out by American billionaire [Bill] Gates with the help of Russian SS-18 missiles (RS-20). The cost of the project is about $5 billion. This system can be used both for military purposes and for information confrontation. A large number of satellites (more than 300) will be able to irradiate any point on earth simultaneously from at least two satellites. The United States now has 420 satellites in orbit and plays a dominant role in the information space on the planet.

So, to protect social objects (individuals, societies, countries) from dangerous informational influences, a system of informational-psychological security as a component of national security must be established.

Targets of informational-psychological influence are as follows:

- the information and psychological environment of a society, which is part of the information environment of the world and uses information, information resources, [and] information infrastructure that can influence the psyches and behavior of people
- informational resources (spiritual, cultural, historical, national values, traditions, etc.)

- the system of formation of social consciousness (worldview, political views, spiritual values)
- the system of forming public opinion
- the system of policymaking
- the psyche and behavior of an individual.

Therefore, informational-psychological security should include the following components:

- protection of the psyche of the population, social groups, military personnel, [and] citizens from destructive informational-psychological influences
- countermeasures against attempts by hostile political forces to manipulate the information perception of the population and the military to weaken the enemy's defense
- upholding national interests, goals, and values in the information space (global, regional, subregional, national)
- constant monitoring of society's attitude toward the most-important matters of national security (diagnosing public opinion), mental state of the population, [and] military personnel
- counteraction to information expansion in the spiritual and ethical spheres.

Throughout history, information was the target of conflict. Information confrontation was waged in almost all wars. For a long time, it comprised intelligence and counterintelligence activities. Since the development of a unified global information space, information confrontation began to undergo fundamental quantitative and qualitative changes. The modern technological revolution transformed the information support of human activities. Mass information—print, audio, video, and other messages for an unlimited number of people—appeared. The means for their rapid dissemination were created.

Information confrontation in the military domain comprises three components.

The *first* [component] is a set of measures to obtain information about the adversary and conditions of information confrontation, collect information about one's own troops, [and] process information and exchange with bodies (points) of command with the purpose of organization and carrying out military actions. The information must be accurate, precise, and complete, and the process of informing [must be] selective and timely. It is logical to call the performance of these [measures] the information support of the military and weapons management.

The *second* [component of information confrontation] is informational influence. It includes measures to block extraction, processing, and exchange of information to disinform.

The *third* [component of information confrontation] includes measures of information protection, such as unlocking the information needed to execute the tasks of command and blocking disinformation [that] infiltrates and [is] disseminated inside the system of management.

When developing a theory of information confrontation, it should be considered that [information confrontation] must be conducted at the strategic, operational, and tactical levels. The highest state bodies should act mainly at the strategic level, [and] intelligence agencies and army units [should act] at the operational and tactical levels.

The role and place of information confrontation in the system of national security of any given country is gradually increasing. Today, leading world countries (the United States, Japan, France, Germany, etc.) have at their disposal information potential that can enable them to achieve political goals, especially because there are still no international legal norms for information confrontation.

To protect social objects from the negative effects of global information confrontation, a system of informational-psychological support as a component of national security must be created. This system should protect the psyche of the individual, society, and state from negative informational-psychological influence.

Informational-psychological influence is the purposeful production and dissemination of special information that has a direct impact (positive or negative) on the functioning and development of the informational-psychological environment of the society and the psyche and behavior of the population [and] military personnel. Varieties of informational-psychological influence are psychological and propagandistic influences.

Because of the emergence and accelerated development of electronic mass media, the role of public opinion has increased considerably and it began to exert a powerful influence on sociopolitical processes, features of the informational-psychological environment of society, [and] the mental state of military personnel during wars and armed conflicts. Therefore, the system of forming public opinion is one of the main foci of informational-psychological protection. It is also necessary to study the peculiarities of the formation and functioning of public opinion in times of armed conflict, which can serve as a basis for practical steps to ensure the psychological safety of military personnel.

Extensive informatization of the armed forces has created a new situation in the development of military affairs. This is illustrated by the armed conflicts and local wars of recent decades. Analysis [of these conflicts and wars] shows that the course and results of combat of any scale in the modern world are largely determined by the [combatants'] skill [with] using information confrontation.

Analysis of armed conflicts in the second half of the 20th century shows a transfer of efforts in the use of forces and means of informational-psychological influence to an earlier period (from one to two months to several years prior to any military actions); new means and methods of informational-psychological influence (information weapons) were developed.

Information warfare can be used to influence the following: informational-technical systems, informational-analytical systems, informational-technical systems involving humans, informational-analytical systems involving humans, information resources, systems of formation of social consciousness and opinion that are based on the media and propaganda, [and] the psyche of an individual.

In cases where information warfare is directly or indirectly used against the human psyche (or against a societal group), this is informational-psychological influence.

Humanity has become hostage to sophisticated information technologies and careless use [thereof]. It is not hard to imagine the consequences of accidents in the telecommunication systems that control pipelines, energy systems, environmentally hazardous industries, satellite communication systems, [and] air or rail transport.

A special place in the information domain of society is occupied by individual, group, and mass consciousnesses, which are increasingly exposed to aggressive information influence leading to the loss of mental and physical health of citizens, the destruction of the moral norms of society, [and] the destabilization of sociopolitical situations. Therefore, the protection of individual, group, and mass consciousness of citizens from unrighteous informational influences should be the main focus of information security activities.

The domain of individual consciousness. The basis of information security of individual consciousness is the ability of a person to adequately perceive reality [and] his own role in the world, [and] to form opinions on the basis of experience and make adequate decisions.

In this case, the main threat to individual consciousness will be a violation of such ability through covert and incognizant influence on the mental structures of one's subconsciousness (for example, the "25th frame") or consciousness, which will create the possibility of the "forced" change of one's mental reactions and behavior.

Religious sects and groups that preach fanaticism, extremism, and misanthropy pose a significant threat to the individual consciousness of citizens. The rituals of these sects and groups are usually aimed at changing the motivational attitudes of their members that contradict the norms of public morality, [including the] development of stereotypes of antisocial behavior and dangerous psychological treatment.

One of the sources of threats to the individual consciousness, especially to those who are below the poverty line, is aggressive advertising of expensive goods that imposes a focus on joining the "world elite" by any means. Perhaps this is what contributes to the growth of crime, animosity, [and] the involvement of youth in criminal activities.

No less dangerous are representatives of the "occult sciences," who offer services of "warding off the evil eye," "return of loved ones," [or] "enchantment with the use of blood" with a "two hundred percent guarantee." These "specialists" often do not have appropriate medical education, so their activities can also be harmful to mental health. Mental health disorders [and] actions that are detrimental to society and state interest can be the consequences of realization of this threat.

Domain of group consciousness. Information security of the group consciousness is based on the common interests of the group, which form the purpose of its creation, [its] rules of conduct that are accepted and internalized by the members of the group and embodied in their individual consciousness, the [group's] ability to meet these interests and achieve the goals, and [its] readiness to counter existing threats. Protection of the right of association and conditions for its realization are necessary for the development of civil society.

Threats to group consciousness can manifest themselves in the form of illegal information influences of other groups—[for example] civil or state organizations with the purpose of eliminating the common interests of the group—to create difficulties to realization of these interests, to discredit group members, [and] to impose psychological pressure on them.

The sources of threats to the group consciousness are unfair competition from the other groups, criminal "redistribution" of property, [and] confrontational relations among religious denominations and ethnic groups. The consequence of the manifestation of these threats may be the breakup of the group [and] disruption of interaction with other groups [and] civil and government organizations.

The domain of mass consciousness. Information security of the mass consciousness is based on traditional and dynamic components.

The traditional component of mass consciousness is formed by a set of

- common interests of large masses of citizens (social groups or classes, national entities, nations, peoples, or the population of the countries, which can be considered group associations)
- cultural, spiritual, and ethical values acknowledged by [citizens, and] customs, which have formed and establish socially acceptable rules of conduct and lifestyle
- the readiness of these bodies to counteract threats to these interests, values, and customs.

Threats to such a component of mass consciousness are realized in the form of the forcible implantation of interests, values, and customs that are alien to this association. Their manifestation can cause destruction of the spiritual health of the association and its stable social relations, and [lead to] its disintegration.

The dynamic component of the mass consciousness of human association is formed through reflections of information about socially important events into the traditional component and causes a corresponding emotional assessment of these events.

Threats to the dynamic component of mass consciousness include distortion of information about events and manipulation of such information to form the needed emotional assessment of events. These threats manifest themselves in a violation of an adequate perception of reality that corresponds to the traditional component of the mass consciousness of human association. Inadequacy of perception, depending on the persistent stereotypes of the association's behavior, may manifest itself in a form of social apathy or aggression toward the outside world.

Sources of threats to the mass consciousness in the context of economic crisis and an uninformed civil society are the uncertainty of spiritual values [that are] protected by the state, inconsistency of actions of government and civil society organizations, lack of success in conducting economic and sociopolitical reforms, [and] the social orientation of these reforms.

The information security system provides for the development of a corresponding system of countering the previously mentioned threats. In the general case, four main components of such system can be identified: regulatory, organizational, technological, and human resources.

The regulatory component must ensure the formation and enhancement of a system of legal norms to counteract threats to information security and mechanisms for the realization [of those norms]. [This component] is formed by a set of legal acts [and] other regulations that govern relations in the field of identifying threats to the security of individual, group, and mass consciousness and counteracting these threats, which ensures the realization of constitutional rights and freedoms, their legal restrictions, the protection of the mental health of citizens, [and] preservation of the social order.

The organizational component of the information security system has to establish the structure of civil society organizations and state agencies [that are] involved in the realization of legal norms in this domain and the relationship between them, as well as [the relationship] between these organizations and agencies on the one hand and the population on the other. The corresponding structures of civil society should be the most important parts of the organizational component of the system.

The organizational component is an important part of the overall system of information security, the configuration of which should be indicated in the doctrine of information security of a country. The information security system should be based on close cooperation among the head of state, legislative, executive, and judicial authorities, as well as civil society organizations that perform legally allowed activities in this domain.

The technological component of this system must provide the possibility of free and secure exchange of information among citizens, members of groups, [and] bodies, as well as the prevention of unlawful information influence on them; [the] timely detection of threats to the information security of the individual, society, and the state; [and an] assessment of possible and caused damage to this security and the organization of effectively countering such threats.

The human resources component must ensure the development and maintenance of the human potential of society and the state that is necessary for the effective functioning of the information security system.

The following important issues of information and psychological security of the state—which demand an urgent solution—are also worth highlighting.

First, public policymaking in this domain because of the specificity of the object and the subject of security: To a large extent, social relations—which arise with favorable conditions for the formation and development of the spiritual domain of society and ensuring the security of these processes—must be regulated by society independently through the creation and maintenance of the criteria of morality, acceptable stereotypes of behavior, and mechanisms of social influence on violators of these established rules. The state, through civil law, must ensure the prevention of the most socially dangerous actions in this domain. Errors in separating these groups of relations lead to the insufficient effectiveness of legal protection of the individual, society, and the state; discreditation of the government; and the lack of proper attention to institutions needed to solve the problem.

Second, enhancement of the system of media, which has the most significant impact on individual, group, and mass consciousness: On the one hand, there are no sufficiently effective mechanisms for society to influence the media to protect the interests of social morality, the mental health of citizens, [and] their serenity; on the other hand, the government is slowly working on the establishment of open information resources that provide citizens with the opportunity to obtain reliable and complete information about the most important events of social life [and] government actions to counteract the existing threats.

Third, there are significant difficulties in assessing the loss of mental health of citizens. The[se difficulties] are based on the lack of sufficient technological tools for solving this problem; the necessary methodological apparatus of recognizing and fixing the psychological characteristics of a particular person, the dynamics of their changes, [and] identifying the causes of negative trends. This is especially important for conducting forensic examinations on the facts of illegal influence on the psyche of an individual and the development of comprehensive techniques and tools to increase the resilience of a psyche to negative information influences, including through channels of mass information.

A separate aspect of information security of a state is the establishment of a training system to conduct preventative work in this domain, [as well as] examinations and regulatory measures to create legal and technological support.

CHAPTER FIVE

Information Resources and Information Confrontation

The wars of the future will certainly include such an important element as information confrontation.[1] The information resource will become one of the components of the state's strategic offensive and defensive forces. Intelligence will also be subject to significant development. It will transition from a traditional type of support for combat operations [as it has been] in past generations of wars into an active operational type of force, and will become one of the strike components of high-precision weapons and defense systems.

The Increasing Role of Information Confrontation in Contactless Wars

The information-scientific and technological revolution is one of the most important mechanisms for the formation of modern views on the conduct of combat activities, which are now also undergoing a stage of the [development] of global information systems.

Despite the fact that many elements of the confrontations of the previous generation of wars are still active and persist, we are now witnessing a sharp leap in informatization and automation of command and control of troops and weapons. We see a rapid process of automation of all levels of the organizational structure of the armed forces; however, during this transitional

1 [Slipchenko, 2013.]

period, information confrontation remains a support element for all other types of warfare.

Subsequently, after the end of the transitional period, information confrontation will gradually go beyond the limits of [being a] support element and become a combat [element]; that is, it will acquire an independent character among many other forms and methods of warfare. However, unlike high-precision offensive weapons that strike a specific, specially selected important target or its critical node, information weapons will be system-destructive; that is, [they will] incapacitate entire combat, economic, or social systems. Superiority over the enemy will be achieved through [having] an advantage in [1] collecting various types of information, [2] mobility, [3] reaction speed, [4] precise fires and information effects in real time against numerous targets of its economy, [and (5)] military facilities, and at the lowest possible risk to one's own forces and means. It is quite obvious that to prepare for the conduct of contactless wars, a sovereign state will need to transition from an industrial to an informational society. And regardless of the size of the territory, it will be necessary to have sufficient information resources, especially for space assets, to ensure the ability to carry out comprehensive support for each high-precision offensive or defensive weapon.

Information superiority in contactless wars is likely to be achieved through the following:

- [dominating in terms] of space systems and means of reconnaissance, warning, navigation, meteorology, control, and communication in the information space
- [having] an advantage in the number of high-precision missiles, reconnaissance, and strike combat systems, including ground, sea, air, and space-based components and the ability to maneuver these forces, assets, and fires continuously
- quickly integrating combat software programs with high-precision missiles of various types
- having the ability to employ high-precision weapons of various types on a mass scale and for a prolonged period
- targeting all-around material and technical support for reconnaissance and strike combat systems

- reliably protecting information on high-precision offensive and defensive forces and means on land, in the air, in space, and at sea.

It is quite obvious that information superiority in contactless wars will be the tool that will provide the attacking side with the opportunity to use heterogeneous forces and means in an air-space-sea strike operation, to increase the protection of the delivery systems of its high-precision weapons and the high-precision weapons themselves, to employ reconnaissance and strike combat systems in battle in accordance with the objectives, and to provide agile logistics to support these activities.

Maximum integration of mechanisms for receiving, processing, and analyzing information from multifunctional offensive and defensive systems [that are] based on automatic control systems and a significant reduction in the number of levels of command and control of combat systems, forces, and means will be required. It will also require reliable protection of individual offensive and defensive elements of high-precision combat systems and the strategic system as a whole from all types of modern information effects.

The information resource of high-precision weapons will have to have a full set of software tools and measures of both active and passive protection against attacks on its information systems and its active and passive effects against all existing and future enemy air defense and missile defense systems. It is likely that precision weapons will be linked to air, space, and naval target reconnaissance assets and a joint passive detection and guidance system.

It becomes quite clear that a continuous "information confrontation" will be one of the indispensable attributes in contactless wars. It is true that some Russian experts, apparently referring to certain Western sources, are trying to assert that not an "information confrontation" but an "information war" will be waged. However, the concept of "war" in this situation does not fit at all, because it refers to a more complex sociopolitical phenomenon. War is a special state of society associated with a sharp change in relations between states, peoples, [and] social groups and characterized by [the use of] armed force to achieve political, economic, and other goals. It is a confrontation between social systems, classes, nations, [and] states with the use of diplomatic, political, informational, psychological, financial, and economic methods; armed forces; and many other forms and methods of warfare to achieve strategic and political goals. [Here,] we are talking specifically about information confrontation.

It seems that, in the next 20–40 years, we should not expect the appearance of the next—seventh—generation of wars in which information confrontation will build a system of the type of confrontation that, in fact, is becoming the basis of "information wars." If such wars arise in the future, then they, of course, will be waged in the global information space and mainly by informational means. This next generation of wars may arise, most likely, in the very distant future; that is, not earlier than 50 years from now. But before that, it will be information confrontation. Moreover, it is already clear that such a future war will not be fought only by informational means and will not be directed against a specific enemy or a specific country. Most likely, we should expect the development of a system of different types of forces and means [that is] capable of disrupting the process of normal operation of the information space and global information resources [and] the life support environment of everyone on earth. The wars of subsequent generations will certainly go beyond the operational and even strategic scale and will immediately become global in scale. Using information networks and resources, a global aggressor can provoke man-made disasters in large economic areas, regions, and parts of the world. It is possible that, after 2050, environmental weapons can also be developed with the purpose of targeting the mineral and biological resources of countries, individual local areas of the biosphere (atmosphere, hydrosphere, lithosphere), [and] climate resources in some parts of the earth. It is important to note that in the wars of subsequent generations, starting with the sixth, an individual will not be the main target. [The individual] will be affected indirectly, through the defeat of other structures and systems that sustain his life.

As far as contactless wars are concerned, "information confrontation" or "information warfare" is a completely legitimate concept of confrontation in contactless wars and expresses the struggle between opposing sides for superiority in the quantity, quality, and speed of collecting, analyzing, and using information.

FIGURE 5.1
Distributed Denial of Service Attack Architecture

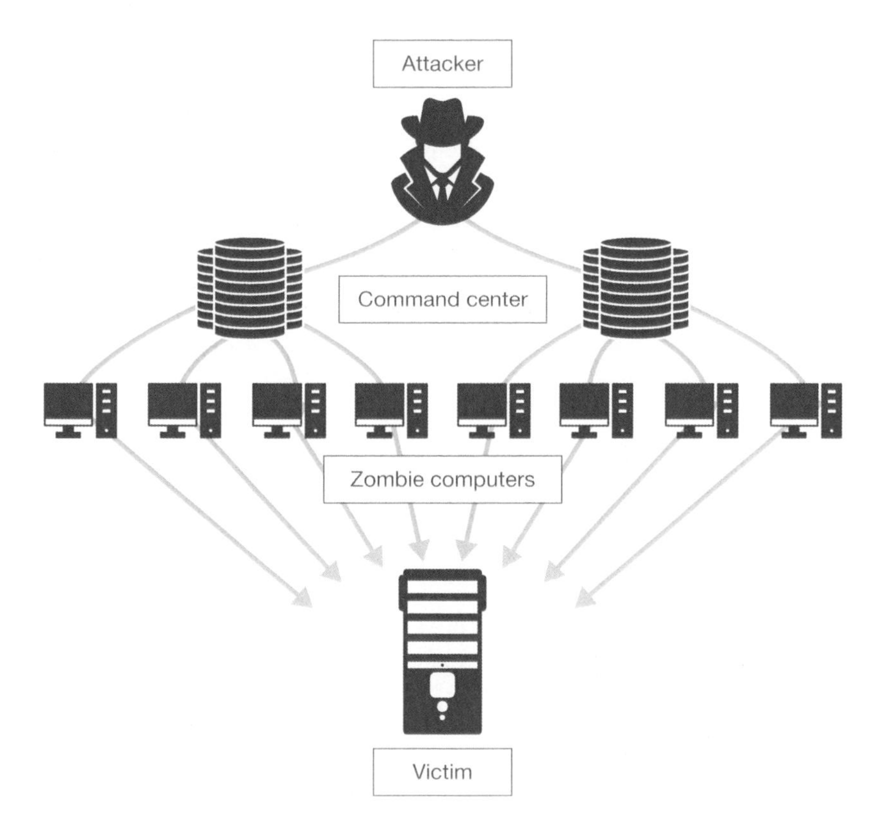

It is clear that this type of future confrontation, like other types [of confrontation], already has two clearly defined components: defensive and offensive or strike.

[The] defensive [component] consists of protecting one's own information infrastructure and information itself from the enemy and ensuring the security of one's own information resources.

[The] offensive [component] consists of disorganizing or destroying the enemy's information infrastructures and disrupting their process of operational control over their forces and means.

Such tools and methods for ensuring the security of one's own information systems and resources as operational and strategic camouflage, physical protection of information infrastructure facilities, counter-disinformation, [and] electronic warfare can be considered defensive.

Such methods of warfare as strategic camouflage, disinformation, electronic warfare, physical destruction and annihilation of information infrastructure targets, "attacks" on the enemy's computer networks, "information effects," "information intrusion," or "information aggression" can likely be employed as offensive components of information confrontation in contactless wars. All of this can be realized across a wide spectrum of specially developed influence means: computer viruses and logic bombs, previously implanted into information systems and networks and triggered by a specific command. "Psychological attacks" or "psychological aggression" in the form of holographic images at high altitude in the sky, for example, [or] information of a religious nature, aimed directly at the enemy's personnel or at its population, can also prove useful here.

Because the movement toward contactless wars is already clearly marked, we should expect that the role of information confrontation will significantly increase in the following directions:

- in the fight against control systems of strategic offensive and strategic defensive forces of various levels
- in the fight between offensive and defensive methods between opposing sides
- in the creation of a complex information and jamming environment in the air and space domains in combat operations and throughout the theater of war (military operations)
- in imposing on the enemy one's own rules of conducting military operations because of the ability to provide information support for mass high-precision missile strikes in all directions
- in relying on information assurance of military and technical superiority.

Information confrontation is multifaceted [and] multifactorial and uses systematic methods of information effects and actions. It developed considerably after the creation of modern methods of military systemology. Using these methods, one can quickly find the most-vulnerable spots in the control systems, communications, computer support, reconnaissance, and all-around support for enemy combat operations and, by disabling them, significantly increase the effectiveness of one's own activities in other types of warfare. Information assets will always be the critical links of the enemy's control system, the suppression, destruction, or annihilation of which will lead to an immediate decrease in their ability to control combat systems, forces, and means and, therefore, to deliver mass high-precision missile strikes against targets of economic potential.

Electronic suppression will likely remain the most important component [of information confrontation]. It is already one of the most effective types of combat support in modern warfare. Electronic suppression will shed the status of combat support and become an independent type of confrontation in contactless wars.

A storm or even a hurricane jamming the electronic environment can be created over the territory of an enemy who is not prepared for contactless wars, against which the aggressor will be delivering continuous mass strikes with high-precision cruise missiles and other missiles. As a result, absolutely all radio-electronic assets on the ground, on the water, underwater, in the air, and in space will be blocked. Those assets that continue to function and emit electromagnetic energy will be immediately destroyed. In the interests of information confrontation, the creation of global combat information-strike systems that are capable of controlling the state and operation of armed forces, groupings, offensive and defensive forces, and means of any opponent and, taking appropriate measures, [capable of] reducing the effectiveness of their actions, will likely be a critical strategic goal of industrialized countries.

The rise in interest in information confrontation in the wars of the future is not an accident; that is, this [rise in interest] is linked to the fact that information is becoming the same type of weapon as rockets, bombs, torpedoes, etc. Today, it is clear that information confrontation will become a factor that will have significant impact on the beginning, conduct, and outcome of the wars of the future.

Thus, it should be noted once again that information confrontation in contactless wars should be understood as a new, strategic form of warfare between opposing sides, in which special methods and means are used that affect the enemy's information environment and protect their own in the interests of achieving the strategic goals of the war. However, it is worth noting that information confrontation as a form of support for military operations almost never ceased in all past generations of wars—it is still ongoing. [O]pposing sides have always sought to control the enemy's information appropriately, not only in wartime, but also in peacetime.

The possession of information resources in wars of the future is becoming the same indispensable attribute as the possession of forces and means, weapons, ammunition, transport, etc. was in past wars. Winning information confrontation in contactless wars of the future will actually lead to the achievement of the strategic and political goals of wars, which will be akin to the defeat of the enemy's armed forces, the seizure of the enemy's territory, the destruction of its economic potential, and the overthrow of the political system.

The goals, objectives, forces, and means of information confrontation are the basis for constructing its definition, and, consequently, the structure of its scientific theory.

In its most general form, the main goal of information confrontation (information warfare), as has already been shown, is to maintain the required level of one's own information security and reduce the enemy's level of [information] security. This goal can be achieved by solving several interrelated tasks, the most important of which will be the destruction of the enemy's information resources and domain and the preservation of one's own information resources and domain.

The first experience of conducting information confrontation as one of the components of military confrontation on an operational scale was acquired in the Persian Gulf war in 1991. Then, the multinational force, using radio-electronic and fires countermeasures, carried out a blockade of practically the entire information system—including the military system—of Iraq. This success not only inspired the multinational forces to understand the role of information confrontation, but also forced them to think about how to deal with a situation where they face the same type of confrontation.

Analytical studies and experiments were carried out in the United States under the leadership of the information security agency of the [U.S.] Department of Defense, which showed that the degree of vulnerability of computer systems and databases of the U.S. military departments is quite high. Apparently, it is not difficult to penetrate the Pentagon's brain center because it has so many different connections to other information systems both inside and outside the state. At present, it is quite easy to disrupt the operation of the information networks of an industrially developed state, not only through traditional communication channels, radio, television, [and] mass media, but also through the internet.

All attacks on the websites followed one simple scheme. Gigantic volumes of empty requests were sent to their servers [and] the servers could not process them and "froze" for several hours (DDoS [distributed denial of service] attack). It should be noted that server hacking did not occur, [and] the security system was not compromised anywhere, but the United States assessed it as "cyber terrorism" for the first time. The importance of internet networks in developed countries is already so great that the slightest encroachment on their inviolability is considered a vital threat to the security of the country. Each new detection of hacking or blocking of internet networks indicates the vulnerability of even the most advanced technology. However, software developers have yet to demonstrate a commitment to protecting the internet.

We should expect that it will be possible to conduct psychological influence against the enemy through the same channels without any witnesses, and to warn one's own state in advance about the threat to [the state's] national interests. Access to a global computer network makes it possible to transmit the necessary information to any region of the world and to perform many tasks associated with information confrontation. It is possible that cyber warfare can develop independently within the information confrontation framework, during which powerful information strikes will be delivered against the enemy's integrated computer systems. Information intrusion can be carried out through the internet to disrupt the enemy's life-sustaining systems, electricity, gas, and water supply; paralyze the communication system and transportation; disrupt financial transactions; etc.

Thus, information confrontation is becoming the most important element of modern wars and, clearly, of future generations [of war] as well. In the future, we should expect the forces and means to use artificial intelligence in this confrontation. Information confrontation will acquire significant independence and become an integral element of all other forms of warfare, owing to the ability to help accomplish operational and strategic tasks successfully in the wars of future generations.

Between 2030 and 2050, we should expect significant breakthroughs in the field of artificial intelligence that likely will find widespread application both in offensive and defensive weapons systems, as well as in the forces and means of electronic warfare. The first work in this area began in the 1960s, but at the beginning of this century we should expect the appearance of fundamentally new electronic models of intelligence. They will probably be built like neural networks in the human brain and will be able to process all incoming information simultaneously and, most importantly, they will be able to learn. Artificial intelligence will find widespread application, first of all, in homing warheads for high-precision intercontinental and cruise missiles, as well as in missile defense systems and cruise missile defense systems.

Intelligence in Contactless Wars

A special role in wars and armed warfare of the new generation will belong to intelligence, whose functions will widely encompass all spatial domains: space, air, land, [and] sea. It will also penetrate computer software of various systems and their networks; networks of telecommunication and radio navigation systems; systems for command and control of troops and weapons, energy, transport, mass media, [and] financial flows; etc. The intelligence of the most-developed countries, benefiting from the results of the continuously ongoing informational, scientific, and technological revolution, will certainly become global, integrated with the help of all forces and means available to states, and will be conducted and documented continuously through monitoring. Space, air, sea, and ground forces and assets will be widely used for the conduct of reconnaissance. It will be necessary to monitor continuously and in detail practically all the territories around the globe and the seas and oceans that surround them, the airspace, the state of the strategic offensive and defensive forces of the countries that own them, all movements of troops (forces), and the limits of global theaters of war (military activities). This is because previously developed reconnaissance and strike combat systems that will be able to carry out air, space, and sea strikes on a strategic scale in a contactless way against any country in any region of the world without prior buildup of forces and means in those areas can, quite clearly, begin a new-generation war. Such a war likely is to be controlled directly from the territory of the aggressor state.

High-precision weapons and weapons based on new physical principles will be entrusted with those most important tasks that only large groupings of manpower, supported by aviation and the fleet, accomplished in the wars of past generations, and will require necessary instrumentally accurate intelligence information about each target planned for destruction. It is quite clear that there will be an urgent need for a variety of automatic and automated reconnaissance and information systems of various basing, which will include electronic navigation support with the global positioning system of coordinates, reconnaissance, and control, as well as forces and means of electronic warfare.

Space reconnaissance assets of the future, quite clearly, will become the main source of information for planning, organizing, and conducting combat operations. Radio engineering, radar, photo, television, infrared, and radiation reconnaissance will be constantly and widely conducted from outer space and will continuously provide necessary information in real time. Space assets will support navigation of land, air, sea, and, subsequently, space-based high-precision cruise missiles toward their targets. It is likely that these missiles will be equipped with global space navigation system receivers, which will make it possible to deliver high-precision strikes in radio silence against targets anywhere in the world.

For example, American "Macros" reconnaissance satellites were widely tested in the Persian Gulf wars (1991 and 1996) and in Yugoslavia (1999), transmitting a radar image of the combat area from space. Their use made it possible to reveal the buildup of ground forces, air defense systems, and military and economic targets of the countries under attack with a sufficient degree of accuracy under any meteorological conditions and in various geographic regions.

Already, the quality of orbital clusters of such reconnaissance spacecraft in a number of industrialized countries has significantly increased compared with that in 2001 [through] 2010. Their capabilities have also significantly increased

- by at least five to seven times in simultaneously detecting a growing number of targets in any region of the world

- by eight to ten times in the accuracy of detecting the target coordinates and their critical nodes.

At the same time, the time for reconnaissance and delivery of information to high-precision offensive weapons has decreased by five to seven times.

We can expect widespread use of the most-advanced information technologies—based on automated design systems and computer networks capable of processing almost all intelligence information available to the state and providing the necessary data—first and foremost for strategic offensive and defensive forces. We should expect the networks of the main local area [where the conflict is centered], scientific and research organizations, and planning agencies of state security bodies to unify into a global network and to integrate all available databases related to defense capabilities.

High-precision cruise and other conventional land, space, air, and sea-based missiles in existence and in development in the leading countries of the world will likely be used only under conditions of information superiority. This will require using intelligence, informatics, and communications to obtain the most complete, accurate, timely, and secure intelligence information, quickly allowing one to respond correctly to any military conflict in any region of the world to master the situation immediately and make the necessary decisions. It is likely that completely new global military intelligence, command, and communications systems can be developed for this purpose in the period from 2020 to 2030. During this period, the most developed countries will, most likely, create space communication and information networks, practically covering all spheres of armed warfare around the globe. Intelligence information will likely receive its own channels and will be automatically delivered to the interested command and control bodies across all levels of authority in the shortest possible [amount of] time. We should expect the simultaneous development of systems that prevent the enemy from collecting information to control their own combat systems and weapons.

International space law does not yet have a treaty establishing the boundary—located at an altitude of approximately 60–100 km above the Earth's surface—between air and outer space. In this regard, near-Earth space will not only retain great military importance as a support domain for military operations but also likely will become a primary theater of war (military operations), where fierce armed conflict can take place and from where nonnuclear weapons can be launched, including in relation to objects and targets on any continent, in any region of the world. It will be absolutely necessary for every country preparing for or already ready to conduct contactless wars to have complete control over the near-Earth and interplanetary space.

Control of all reconnaissance combat systems and forces and means will most likely be carried out from command posts in space and in the air, or from protected command posts on the ground. We should expect a significant increase in the number of command-and-control aircraft and long-range non-radar detection. Information exchange will likely be carried out between all ranks and files of command using automatic or automated systems deployed on air and space vehicles.

Reconnaissance and strike combat systems (RSCS) that are based on manned and unmanned space reconnaissance and information systems, as well as on ground, sea, air, and space delivery systems and on high-precision offensive weapons, will find wide application in contactless wars. [Such systems] will be capable of detecting and delivering effective strikes against stationary radio signal and heat-emitting military and economic targets; ground components of air, space, and naval defense systems; and radar-emitting sources deep in the enemy's territory. This will radically change the substance and nature of war. In this contactless war, it is not the troops (forces) that collide but reconnaissance offensive and defensive combat systems. Their capabilities are characterized not by the quantitative and qualitative superiority of one side over the other, but by structural and organizational factors; the unity and effectiveness of control; [and] the operational quality of intelligence systems, communications, navigation, and other links in the comprehensive support of military operations.

It should be emphasized that, after 2020, such RSCS with space-based delivery systems can be developed in advance and covertly as dual-use systems, and all of them can aim at specific and critical static civil and military targets of potential adversaries in any country in the world in advance or at the necessary time.

Reconnaissance satellites are expected to have exceptionally high performance. The latest radar reconnaissance satellites will likely enter military service and will be able to obtain images of terrain with a resolution of several tens of centimeters in the dark and in conditions of dense clouds. The capabilities of optical reconnaissance satellites can increase significantly. Their resolution will reach ten to 15 cm, which will provide an uninterrupted, detailed view of the entire surface of the earth and all economic targets of countries within one day in the daytime. Information from space reconnaissance assets will be transmitted via communication relay satellites to ground control centers, and from there images and intelligence data will be sent directly to ground, air, and naval command posts of RSCS and assets and will become the basis for planning mass high-precision strikes during strategic air, space, and naval offensive operations.

It is possible that, in some economically developed countries, new space-based missile warning systems—which will be able to detect not only the launch of land-based and sea-based intercontinental high-precision ballistic missiles, but also missile warheads in the middle and final stages of flight—may be adopted for military service. These same satellites will likely be in stationary orbits and will become the main information component of the anti-missile and anti–cruise-missile defense systems of states. We can expect that space reconnaissance of ground targets will also continue to develop because these assets will be conducting completely new tasks of searching, detecting, identifying, and measuring parameters and identifying critical nodes of enemy stationary economic, infrastructure, and military targets that must be destroyed by high-precision cruise or intercontinental missiles with conventional warheads. It is likely that space, sea, and air-based reconnaissance systems included in the corresponding RSCS will be used to control the concentration of high-precision strikes against critical economic regions, industries, and key enemy economic and military targets. Unmanned reconnaissance aircraft that are less expensive than space assets but highly effective will likely be employed in detecting numerous ground targets and identifying their critical nodes, located deep in the enemy's territory.

It should be noted that, even now, different armed forces and branches of the military in several of the most-developed countries have an extremely large number of different types of reconnaissance forces and means, but all of them are largely fragmented and mainly used collectively. High integration of the numerous intelligence systems will first be necessary for their automated distribution across a global information grid in future wars and military conflicts. There will be a need for a serious integration of departmental intelligence systems within the state, as well as space, air, sea, and ground-based intelligence systems within the state's armed forces. Increased flexibility and universal use of reconnaissance assets is inevitable, as well as the design and development of the latest reconnaissance assets, complementing existing assets with new information capabilities.

Thus, already in the first ten years of the new century, the most-developed states that are taking the necessary measures to create virtually new armed forces and weapons will likely have to develop new systems and means of intelligence, which will prepare them to implement the technology to conduct contactless wars.

CHAPTER SIX

Information Confrontation in the Military Sphere

At the heart of the term "information confrontation" lie two concepts: "information" (and ways of using it in the interests of someone else) and "struggle."[1] In their writings, many politicians and generals describe various ways of using information to mislead the enemy, undermine its will to resist, produce panic in its ranks, and generate betrayal. Hence, the essence of information confrontation is the purposeful use of information to achieve political, economic, military, and other goals.

At present, interest in this problem has grown because of the rapid computerization of technology [and] the increased volume of information and speed of its dissemination. At the end of the 20th and beginning of the 21st century, the character of geopolitical conflict between states changed, the process of globalization continued to develop, [and] as a result of this, along with traditional methods of force and means of solving problems in this area, information has been increasingly used. Examples of this kind of combination of methods and means include, in particular, the operations carried out by the United States and its allies in Iraq ("Shock and Awe," "Desert Storm," and others) and the events taking place in Ukraine.[2]

---

[1] [Sayfetdinov, 2014. In the translated English summary of this article, *informatsionnoe protivoborstvo* is translated as *information warfare*. This is reflective of the inconsistent usage of the term in the Russian military-scientific literature. For the sake of consistency, we translate *informatsionnoe protivoborstvo* as *information confrontation* throughout this volume.]

[2] H. Epple, "Rossiia v virtualnoi voine [Russia in a Virtual War]," *Vedomosti [Gazette]*, May 8, 2014.

Back in September 2012, the Chairman of the Joint Chiefs of Staff of the Armed Forces of the United States, General M. Dempsey, signed the "Capstone Concept for Joint Operations," in which forces and means of information confrontation are assigned an important role. This document also specifies that a prospective form of employment of the armed forces of the United States and its allies will be globally integrated operations, the basis of which are operations of special operations forces and cyber operations forces of the armed forces of the United States, conducted simultaneously with or separately from general purpose operations.

The forces of information confrontation are supposed to be tasked with [1] disrupting information links between the military and state administration; [2] reducing their ability to obtain reliable information through space reconnaissance assets, missile warning systems, and control of outer space by affecting mass and individual consciousness; [and (3)] making decisions to reduce opportunities for realizing the combat potential of the armed forces of the Russian Federation. Therefore, the purposeful use of information resources is becoming a decisive factor, in many respects, not only determinative of victory or defeat, but also capable of preventing armed conflict.

The experience of preparing for and carrying out two military campaigns to restore the constitutional order in the Chechen Republic showed that [Russia's] military command did not pay enough attention to the issue of information confrontation. In connection with this, the state and military administrations of our country need to take the most serious approach to this very urgent problem, including by improving the regulatory framework. The military-scientific community must comprehend the practice of information confrontation in both peacetime and wartime, during the preparation for and conduct of military operations by the armed forces of the United States and their allies against Iraq and Afghanistan.

It should be noted that when studying the problem of information confrontation, it is important to clearly define (i.e., clarify)

- basic concepts, terms, and definitions
- the goal and objectives of information confrontation in general, and above all in the military sphere; [and] principles of achieving goals in peacetime and wartime

- possible forces and methods involved in solving the tasks at hand
- effective forms and methods of information confrontation.

It is also necessary to develop requirements for promising means, [provide] recommendations on the organization of planning and management of information confrontation, and provide answers to other questions (including the question of how the theory of information confrontation was created) and prepare a temporary "Manual on Information Confrontation."

In recent years, the question of information confrontation has received sufficient attention on the part of scholars and the military leadership, as evidenced by research that has solved certain problems and tasks; some aspects of this problem are reflected in governing documents. Therefore, building on existing expertise and without claiming their authorship, let us try to formulate some of the basic provisions of information confrontation.

The main goal of information confrontation in the military sphere, in our opinion, is to achieve and maintain information superiority over the enemy's armed forces and to create favorable conditions for the preparation and use of our own armed forces.

Information confrontation should be conducted constantly in peacetime, during a period of competition, and in wartime, with all available forces and means, by influencing the information objects of the opposing side and protecting one's own [side] from such influence.

The main principle of achieving [one's] goal is the combined effect of forces and means of information confrontation on the objects of the enemy, in close coordination and cooperation with the actions of troops (forces).

Forces and means of information confrontation should be combined in a single system and applied in a coordinated manner in terms of their goals, tasks, place, and time.

In peacetime, information confrontation should achieve those tasks set by the country's political leadership and be aimed at increasing the effectiveness of political, diplomatic, economic, legal, and military measures to ensure the security of the Russian Federation, primarily to solve the problem of strategic deterrence. In addressing these challenges, bodies of state and military administration, available forces, and means of information confrontation should be involved.

It is advisable to manage information confrontation and implement military and nonmilitary activities centrally from the command post of the highest echelon (from the command post of the General Staff of the Armed Forces or from the newly created national command post for the country's defense) with the involvement of command posts of military districts, fleets, and territorial command posts.

In a period of competition, the system of information confrontation should address the tasks set by the military-political leadership of the country, taking into account the prevailing situation according to previously developed (specified) plans for the organized entry of the country and the armed forces into war and the successful fulfillment of the assigned tasks, while carrying out activities of a military and whole-of-government (nonmilitary) character.

Information confrontation can be controlled from a single command post deployed at the General Staff of the Armed Forces. Federal executive bodies must determine the forces and means for involving the armed forces in information confrontation.

In wartime, information confrontation should solve the tasks set by the country's highest military-political leadership to achieve and maintain information superiority over the enemy to create favorable conditions for successful actions of groupings of troops (forces) as intended.

Management of information confrontation in the preparation and conduct of strategic activities (operations) by the armed forces of the Russian Federation should be carried out by military command and control entities, taking into account information activities of a whole-of-government (nonmilitary) nature and information measures that are carried out by command and control entities, forces, and means of various ministries and departments. In preparing for and conducting operations (combat activities), all information activities aimed at influencing the enemy's information objects and protecting one's own troops (forces) and weapons [and] the population in the area of the military activities (military conflicts) should be centrally coordinated.

The main tasks of information confrontation could be

- monitoring of information sources [and] the identification, assessment, and forecasting of threats to the Russian Federation and the armed forces in the information sphere
- misleading the opposing side about one's own plans and intentions
- the disorganization (destruction) of governmental and military command and control of groupings of enemy troops (forces)
- reducing the psychological stability of enemy personnel and population during the preparation for and conduct of hostilities
- maintaining the stable moral-psychological state of one's own groupings of troops (forces).

An important task in the course of information confrontation is the protection of one's own automated control systems for troops (forces) and weapons and information, information management, and other systems of military and defense significance.

The Military Doctrine of the Russian Federation establishes guidelines for the appearance of potential military dangers and military threats, as well as in connection to military conflicts. At the same time, the military conflict of the future will be characterized by the quick pace of operations, which will demand an overhaul of the whole cycle of functioning and other parameters of military management. Situational assessment, decisionmaking, [the] realization [of this assessment], and performance evaluation should be performed in near–real-time. It should be noted that, in such conditions, the use of forces and means of information warfare by the enemy can significantly complicate the work of one's own system of governmental and military control. Moreover, it is necessary to take into account that information confrontation will also be conducted in the psychological sphere.

In a prospective form of the armed forces, one of the main systems for ensuring the implementation of measures for one's own information security and suppression (defeat) of the information objects of the enemy must be a system of information confrontation. Its structure can include a number of subsystems:

- information technology impact and protection
- hardware and software impact and protection
- intelligence, including signals intelligence
- electronic warfare
- psychological struggle and moral-psychological support.

In this way, the creation and development of a system of information confrontation should ensure the implementation of the entire spectrum of functional tasks that are assigned to it. At the same time, it is very important to correctly determine its place in the unified system of governmental and military administration.

CHAPTER SEVEN

Information Confrontation on the Operational-Tactical Level

Increasingly, the practice of modern armed conflicts, while remaining the criterion of truth, does not confirm the predictions of many military theorists about the contactless or remote nature of armed confrontation in the 21st century.[1] As before, the statement of the outstanding military historian and theorist A. A. Svechin remains relevant, "the various means that the era gives affect the type and way of applying the strategy, but do not create a new one."[2] Taking into account the means that our era has given us—that is, the informatization of the means of armed warfare and their improved capabilities—it is necessary to develop a unified approach to the phenomenon of information confrontation in the preparation and conduct of an operation (battle).

At present, in the Russian armed forces and most armies of technologically developed countries of the world, information confrontation is understood as a type of warfare between opposing sides, each of which seeks to inflict defeat (damage) on the enemy using information effects on their information space [and] matching or reducing such an effect on their own side. It is deployed in two directions: *informational-technical* is the destruction of information, electronic, [and] computer networks, and unauthorized access to the enemy's information resources (and protection of one's own

---

[1] [Trotsenko, 2016.]

[2] A. A. Svechin, *Strategiya v trudakh voyennykh klassikov [Strategy in the Works of Military Classics]*, Vol. 2, Moscow: Federal Military Edition, 1926, p. 244.

information environment from the enemy); and *informational-psychological* is the impact on the civilian population and enemy armed forces personnel.[3]

However, this understanding of information confrontation addresses the area of strategic and military and state command and control. At the tactical and operational level, information confrontation is limited to individual issues of the organization of command and control and electronic warfare (EW). Attempts to work out the preparation and conduct of information confrontation during training, as a rule, result in fruitless agitation around the same issues of command and control and EW. Adding informational-psychological aspects to these considerations only exacerbates the futility of demagoguery and leads to the loss of time and efforts on the part of operational staff.

At the same time, it follows from the definition of information confrontation that its central idea (goal) is to reduce the effectiveness of command and control of enemy forces and ensure one's own superiority in this area. The consequence of this superiority is the ability to preempt enemy activities, which is, of course, extremely important for the tactical and operational levels of command and control. An obvious connection exists between the effectiveness of command and control, the possibility of achieving superiority in command and control or information superiority, and preempting enemy activities on the one hand, and the effectiveness of reconnaissance, [electronic weapons], security, tactical camouflage, kinetic destruction of elements of the enemy's command and control system, and the use of highly mobile subunits on the other hand.

In this regard, we can confidently assert that information confrontation is effectively present during the preparation and conduct of tactical activities. Thus, ensuring the most effective control of your own troops and reducing the effectiveness of the same for the enemy are the main goals of information confrontation. Achieving the goals of information confrontation can be characterized *by the degree of information superiority* and, as a result, *by the ability to preempt enemy activities early on* or by the ability to prevent the same in defense.

---

[3] Ministry of Defense of the Russian Federation, "Informatsionnaya protivoborstvo [Information Confrontation]," *Voyennyy entsiklopedicheskiy slovar' [Military Encyclopedic Dictionary]*, undated-b.

Thus, information confrontation in the preparation and conduct of tactical actions should be understood as a set of measures that includes the organization and implementation of control; reconnaissance; [electronic weapons]; security; tactical camouflage; kinetic destruction of elements of the enemy's command and control system; the use of highly mobile subunits and certain types of maneuvers coordinated in time, place, and purpose and aimed at achieving superiority in command and control; deception; and preempting enemy activities.

This approach to the concept of information confrontation is not innovative. It was proposed and mathematically described by the seldom-remembered N. N. Badyakin, who developed the headquarters mathematical model "Foresight."[4] He proposed to use the coefficient of the effectiveness of control (K_u) when determining combat potentials and expressed it in the following relationship:

$$K_u = V_k * \frac{(V_s + V_r * K_z) * V_{sv} + V_{pu}}{Z} * \varphi,$$

where

- V_k is a relative indicator of professional readiness and organizational skills of commanders
- V_s is a relative indicator of professional readiness, operation, and composition of headquarters staff
- V_r is reconnaissance capability within the means of the senior commander (air, space, intelligence) to find enemy targets by a specified time and with the required accuracy (measured as fraction of a unit of the total number in a group of enemy targets)
- K_z is the coefficient of defending the targets from being discovered (measured as a fraction of a unit of the enemy's overall reconnaissance capability)

4 N. N. Badyakin, *Instruktsiya po ispol'zovaniyu shtabnoy matematicheskoy modeli «Predvideniye»: Uchebnoye posobiye [Manual on the Use of Headquarters Mathematical Model 'Foresight': Textbook]*, Moscow: Combined Arms Academy of the Armed Forces of the Russian Federation, 2001, pp. 34–35.

- V_{sv} is the communication capability of the command given the possibility of suppression (measured in units of the total number of communication systems)
- V_{pu} is a relative indicator of the staffing at command posts with trained personnel and equipment, given [the staff's] kinetic damage by the opposing side, expressed in fractions of a unit
- φ is the correlation between the time it takes to prepare the environment that requires a new decision or confirmation of the prior decision, and the length of the command cycle.

The practice of using the headquarters mathematical model "Foresight," given the various correlations between the opposing sides' troops of the levels of command effectiveness, has shown that achieving superiority in command and control (or information superiority) can help accomplish offensive tasks, given the quantitative-qualitative ratio of 1:1.5 for opposing sides and lower-than-usual losses. Achieving information superiority or preventing the enemy from attaining [information superiority] in [its] defense provides for the ability to repel an offensive by a more powerful enemy.[5] Moreover, Badyakin's formula, despite the large number of coefficients, can be used independently and reflect fairly objectively the correlation of information confrontation capabilities of the opposing sides, used in operational calculations, or "on the knee," while allowing broader assumptions for indicators and coefficients.

The coefficient φ is of particular interest in the proposed approach. Separating the processes of command and control of the opposing sides from the processes of their armed confrontation (the process of mutual and direct defeat) makes it possible to identify two primary ways to achieve information superiority or superiority in command and control.

[The first way to achieve information superiority] is [through a] direct impact on the flow of information processes (acquisition, collection, generalization, analysis of information, [and] its processing and dissemination to the troops in the form of combat orders) through the enemy's reconnais-

[5] In other words, effective command and control of forces, including information resources, can enable successful offensive combat operations, reduce friendly losses, and lead to victory over a more numerous and capable enemy.

FIGURE 7.1
Means to Achieve Information Superiority

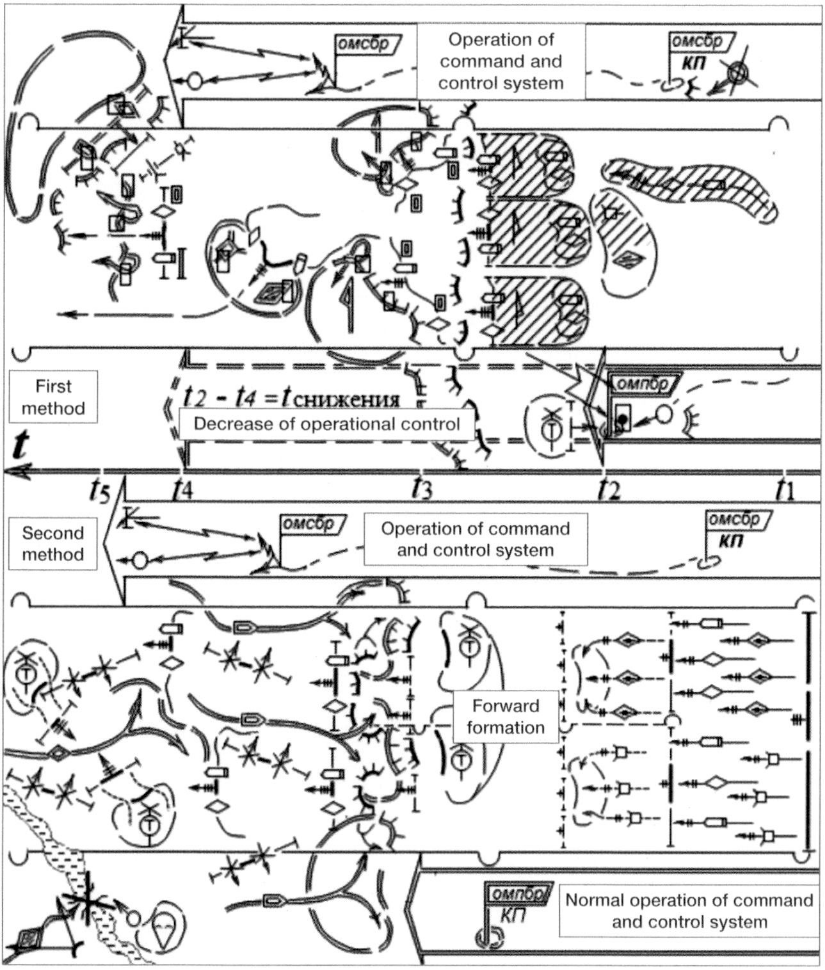

sance and command and control systems and the prevention of a similar impact from the enemy. For this, the most-rational command and control systems for this situation (including the most-rational order for the operation of command and control bodies), reconnaissance, security, EW, and tactical camouflage are being built, [and there is] preparation and employment of kinetic destruction of elements of the enemy command and control system. This method almost completely coincides with the officially accepted one, and [it] can be defined in terms of the *information-technical* component.

[The second way to achieve information superiority] is [through an] impact on the critical nodes of the enemy's armed warfare processes (on the dynamics of changes in the position and state of the enemy troops) across time, using highly mobile subunits and/or certain types of maneuvers at a higher-than-average rate, and prevention of a similar impact from the enemy; or the creation of a situation in which the working effect of a well-functioning enemy command and control system on its own troops lags behind the developing situation, and the prevention of a similar impact from the enemy. This method can be defined as the *preemptive method.*

The simplest example of this situation is the destruction of a bridge by sabotage and reconnaissance forces [DRG] in the path of a second echelon convoy, which makes the calculation for entering the battle incorrectly and requires reassessment; accordingly, the combat command and control cycle doubles and the resources spent on its preparation are meaningless. It is especially important to note that forces with small kinetic and offensive capabilities that are highly maneuverable and capable of concealing their activities can bring appropriate effects against the enemy's critical nodes of armed warfare over time. So, in the [aforementioned] example, finding the DRG in time or transitioning to a defense posture early completely neutralizes its activities because a motorized infantry or tank unit undoubtedly quickly will destroy the detected DRG or maneuver around it, which will preempt the DRG attack. The impact of the DRG will produce the desired effect only when artillery preparation of the second echelon has already begun, the subunit has reached the bridge, and the commanders' response to the situation will not restore the disrupted relationship in time.

Another example is the mass deployment of airborne units to prevent an organized defense by a motorized infantry brigade, followed by a decisive strike while on the move without continuous offensive fires. The conditions for success of the attacking side in this situation are the same as in the first example. The simultaneous defense by units of the motorized infantry brigade will not allow the enemy's commanders to organize an effective fires response in time. At the same time, preemptive detection of assault forces in flight or their activities against the prepared enemy defenses will lead to heavy losses on the attacking side and will not bring success to an offensive on the move. This is precisely the situation that the φ coefficient in Badyakin's formula should reflect, when the correlation of time it takes to create a situation requiring a new decision or clarification of the previous one and the time of the command and control cycle is less than one.

Of course, the effective implementation of information confrontation tasks in practice involves close cooperation of informational-technical methods to achieve information superiority and methods of preemption. Moreover, the effectiveness of the first method is a prerequisite for the successful application of the second, and the experience of information confrontation is likely to put forward a mixed method in the first place. But to understand the essence and definitions of general concepts, we should distinguish three ways of conducting information confrontation.

The idea of preempting the enemy's preparation and conduct in battle and operations is also not new. This idea particularly runs like a thread through the historical observations of Svechin[6] and is discussed in detail in the works of B. H. Liddell-Hart.[7] The latter, broadly understanding indirect actions in preempting the enemy, gives many well-grounded examples from the history of tactics, operational art, and strategy. Examining the actions of German troops in the 1940 campaign in the west, he describes the situation in the following way: "The solution to the problem depended entirely on the time factor. The countermeasures of the French turned out to be unsuccessful, because, as a rule, they were late, lagging behind the rapidly changing situation."[8]

This vision of indirect action certainly coincides with the proposed understanding of preemption. As a rule, Liddell-Hart considers as indirect actions sudden maneuvers to envelop, bypass, and encircle the enemy, while emphasizing the criticality of the sudden use of mobile combat arms. In some cases, preempting the enemy requires bringing troops and forces up to combat readiness. However, pointing out that the ability to deploy amphibi-

---

[6] A. A. Svechin, *Evolyutsiya voyennogo iskusstva [Evolution of Military Science]*, Moscow: Academic Project, 2002, pp. 547–592.

[7] B. H. Liddell-Hart, *Strategiya nepryamykh deystviy [Strategy of Indirect Actions]*, Moscow: Exmo, 2008, pp. 143–167.

[8] Liddell-Hart, 2008, p. 151.

FIGURE 7.2
Activities of German Troops in the 1940 Campaign

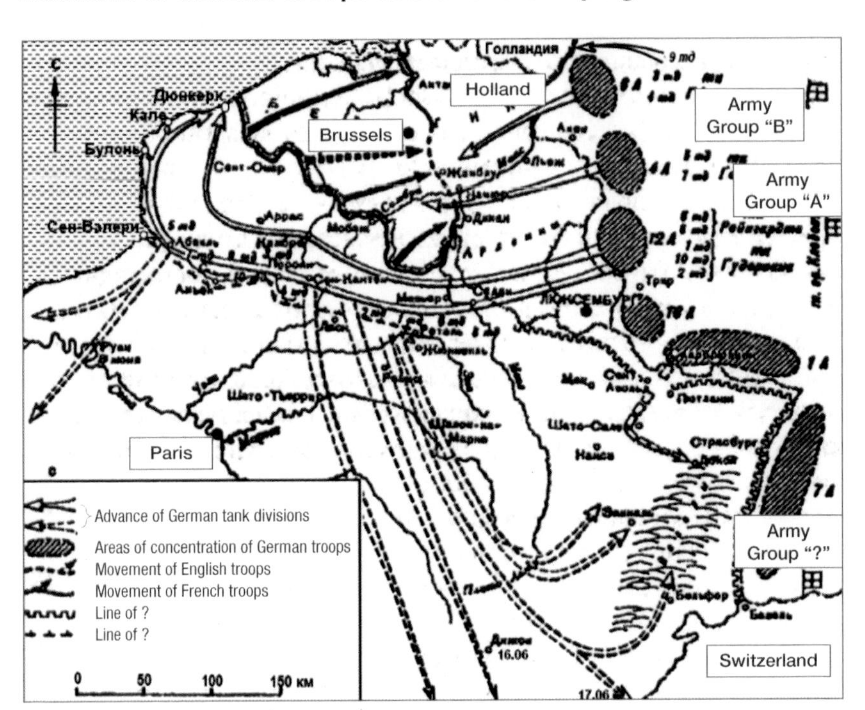

ous and air assault forces by the opposing sides provided special flexibility for the operations of World War II, he does not attach much importance to the methods and conditions of their use.

Meanwhile, it was precisely the emergence of such elements of operational formations (battle order) as amphibious and air assault forces and airborne and special operations forces that provided a number of brilliant examples of preemptive actions during the Second World War. One such example rightfully can be the capture and blockade of bridges over the Albert Channel and the most modern Belgian fort at that time, Eben-Emael, by paratroopers of the 6th Army [of] Reichenau, to ensure the surprise invasion of Belgium by German troops during the 1940 campaign.

Given the characteristics of employing these formations—relatively low fires and offensive capabilities, low defense resilience with high maneuverability, achieving operational secrecy and surprise—logically, their main task should be to change the ratio of the time it takes to create a situation requiring a new decision or clarification of the previous one and the time of the enemy command and control cycle by delivering effects against critical nodes in their armed warfare processes (against dynamic changes in the position and state of enemy troops and forces) across time.

Analysis of the contradictory practice of using airborne and air assault troops and units, marines, and special forces from the information confrontation perspective (in its proposed understanding) makes it possible to identify the systemic reasons for why most of their actions have been unsuccessful; to clarify the goals, conditions, and principles of their use; to determine the right way to employ such new formations as reconnaissance brigades and special forces in the armed forces of all countries; and to optimize the nature and focus of their combat training. However, this is a topic for the next article.

CHAPTER EIGHT

Analysis of Information Wars of the Last Quarter-Century

The history of the development of humanity is a testament to the fact that wars and the use of force to achieve certain goals are an integral part of [history] and cannot, it seems, ever disappear.[1] All development of humanity is associated with wars. Over the past 5,500 years, more than 14,000 big and small wars have been waged, during which more than 3.5 billion people have died from epidemics and starvation—more than half of the current population of the earth. In all the years of the existence of humanity, only about 300 years have been peaceful. Wars today remain the most important factor in world politics.

An analysis of the numerous wars conducted and waged shows that from one war to the next war, there is a continuous process of improving the destructive factors of weapons, the means of their delivery, and the means and forms of their use.

Each type of new weapon appears not out of nowhere, not suddenly, but out of the objective preconditions for the evolution of humanity.

It should be noted that the essence and substance of war, as a rule, are determined by the dominant technological order at a given stage of the development of society.

At present, humanity is entering the era of a new technological paradigm—informational, which is leaving the embryonic phase [of development] and rapidly entering a phase of growth. In turn, this paradigm changes the essence and substance of modern wars; that is, humanity is

---

[1] [Novikov and Golubchikov, 2017.]

entering the era of information wars, which is now, for us, an objective reality. This technological paradigm stimulates the emergence of numerous and varied means and technologies of informational-psychological influence on the individual, collective, and societal consciousness of man to guide his behavior in the right direction.

The emerging hegemony of the West—in the fields of nano-, bio-, information, and telecommunications technologies, energy, [and] science in general—as the basis of a new technological paradigm leads to the dominance, first of all, of the United States and the West in all other spheres of human activity, which creates more-favorable conditions, including for waging information wars.

It should be noted that information and telecommunications technologies have led to the fact that, at present, it is almost impossible for a person to hide from means of influencing his consciousness and subconsciousness. From birth to death, they are with him and act continuously [on him]. It is very difficult to resist the massive, aggressive information-psychological impact, which removes a person from mental equilibrium and leads him to inappropriate behavior. This can be clearly seen from the events currently taking place in Ukraine. This was confirmed by the President of the Russian Federation, V. V. Putin, speaking to reporters in the presidential pool during the BRICS [Brazil, Russia, India, China, and South Africa bloc] summit in Goa, India, on October 16, 2016, having said that the "United States authorities are spying and eavesdropping on everyone."

Experts in information warfare emphasize that, according to a "price-performance" ratio, informational-psychological and informational-technical means and technologies of influencing the enemy significantly exceed the impacts carried out through conventional systems and weaponry.

Information warfare is the continuation of state politics, which consists of a targeted, comprehensive, organized information impact on the information objects of a foreign state through means of information violence to achieve political, economic, national, religious, and other goals with the infliction of minimal loss of life, population, and enemy objects [and] without direct occupation of territory and protecting [the state's own] information objects.

The main goal of information wars is to establish power over the individual and societal consciousness of the individual and society (population) of the "victim" state and to ensure that it fulfills the will of the victor to achieve the goals of the war.

Analysis of the major wars and armed conflicts conducted by the United States and the West in the form of soft (color, velvet) revolutions (information wars) and power [revolutions] since 1991—by changing state structures, heads of state, and governments to the "necessary" [forms]—shows (testifies) that they reached 39 states.

Let us give a brief description of the events that have occurred since 1991, as shown in the table.

The first information war, in the direct sense of the word, was conducted in January–February 1991 by a coalition of states, led by the United States, to "liberate Kuwait from Iraq." The massive use of informational-psychological means and technologies of influencing the Iraqi armed forces and the population through the simultaneous use of conventional weapons and electronic warfare allowed the multinational forces to sustain minimal losses (less than 1,000 people, 148 of them Americans). Iraqi casualties amounted to over 100,000 people.

The same year, as a result of the long-term, integrated effects from foreign states, including informational-psychological [effects], the USSR [Union of Soviet Socialist Republics] split into 15 states. The collapse of the USSR was preceded by a number of external and internal events.

External events—first of all, the dissolution, in June 1991, of the Council for Mutual Economic Assistance, [which had been] established by the decision of an economic conference of representatives of Bulgaria, Hungary, Poland, Romania, the USSR, and Czechoslovakia in January 1949; as well as the collapse, in July 1991, of the Warsaw Pact, which was signed on May 14, 1955, in Warsaw, the capital of Poland. [The Warsaw Pact] included [the following] states: Albania, Bulgaria, Hungary, the GDR [German Democratic Republic], Poland, Romania, the USSR, and Czechoslovakia.

Events within the USSR: December 16–18, 1986, in the city of Alma-Ata [Kazakh Soviet Socialist Republic (SSR)]; July 6–August 7, 1987, the demonstration of the Crimean Tatars in Moscow; February 8–29, 1988, in Sumgait, Azerbaijan SSR; April 4–9, 1989, in Tbilisi, Georgian SSR; July 15–16, 1989, in the capital of Abkhazia, Georgian SSR; June 2–15, 1989, in the Ferghana region of the Uzbek SSR; June 17–18, 1989, in the city of Noviy Uzen, Kazakh SSR; October 22–November 16, 1989, in Moldova; January 13–20, 1990, in

TABLE 8.1
Information Wars Since 1991

	Kuwait	USSR	Yugoslavia	Serbia	Venezuela	Iraq	Georgia	Ukraine
Year	1991		1999	2000	2002	2003		2004
Means of influence used (defeat)	CW	IPMT	CW	IPMT	IPMT	CW	IPMT	IPMT
Goals achieved (cumulative)	+1	+2	+3	+4	—	+5	+6	+7
Goals not achieved (cumulative)	—	—	—	—	+1	—	—	—
Number of countries (cumulative)	2		3	4	5	6	7	8

Table 8.1—Continued

	Kyrgyzstan	Azerbaijan	Uzbekistan	Belarus	South Ossetia	Armenia	Moldova	Iran	Kyrgyzstan	Belarus	Tunisia
Year	2005			2006	2008		2009		2010		2010–2011
Means of influence used (defeat)	IPMT	IPMT	IPMT	IPMT	CW	IPMT	IPMT	IPMT	IPMT	IPMT	IPMT
Goals achieved (cumulative)	+8	—	—	—	—	—	—	—	+9	—	+10
Goals not achieved (cumulative)	—	+2	+3	+4	+5	+6	+7	+8	—	+9	—
Number of countries (cumulative)	9	10	11	12	13	14	15	16	17	18	19

Table 8.1—Continued

	Egypt	Yemen	Libya	Kazakhstan	Syria	Paraguay	Turkey	Egypt	Ukraine	Thailand
Year	2011				2011–today	2012	2013		2013–2014	
Means of influence used (defeat)	IPMT	IPMT	CW	IPMT	CW	IPMT	IPMT	IPMT	IPMT	IPMT
Goals achieved (cumulative)	+11	+12	+13	—	—	+14	+15	+16	+17	—
Goals not achieved (cumulative)	—	—	—	+10	—	—	—	—	—	+11
Number of countries (cumulative)	20	21	22	23	24	25	26	27	28	29

Table 8.1—Continued

	Hong Kong	Venezuela	Macedonia	Armenia	Moldova	Kazakhstan	Armenia	Turkey	Brazil	Iraq
Year	2014	2014–today	2015		2014–2015	2016				2016–today
Means of influence used (defeat)	IPMT	IPMT	IPMT	IPMT	IPMT	IPMT	IPMT	IPMT	IPMT	CW
Goals achieved (cumulative)	—	—	+18	+19	—	—	+20	—	+21	—
Goals not achieved (cumulative)	+12	—	—	—	+13	+14	—	+15	—	—
Number of countries (cumulative)	30	31	32	33	34	35	36	37	38	39

NOTE: CW = conventional warfare; IPMT = informational-psychological means and technologies of influence on an individual, society, [and] population.

the city of Baku, Azerbaijan CCP; January 13, 1991, in the city of Vilnius, Lithuanian SSR; and January 20, 1991, in the city of Riga, Latvian SSR.

From 1991 to 1999 there was a break in the conduct of wars. The United States and the West needed to "digest" the results of the collapse of the USSR.

In 1999–2000, Yugoslavia was destroyed, and a change of leadership in Serbia was carried out. For a long time, Russia was driven out of Europe.

It became clear to the United States and the West that the achievement of political, economic, territorial, and other goals could be achieved with minimal losses, without occupying the territory of the "victim" state, because of the mass application of informational-psychological methods and technologies of influence individually, on a specific person, the population of the state as a whole, and its leaders.

From 2003 through 2011, information wars swept across the states that emerged after the collapse of the USSR (Georgia—2003, Ukraine—2004, Kyrgyzstan—2005 and 2010, Azerbaijan and Uzbekistan—2005, Belarus—2006 and 2010, Moldova—2009, [and] Kazakhstan—2011), with the aim of [these states'] final separation from Russia and the impossibility of restoring the USSR in any form.

From 2010 through 2011, information wars took place in the states of North Africa (Tunisia, Yemen, Egypt, Libya) and Syria ("Arab Spring") with the aim of ousting [the presence of] Russia and China from them, monopolizing carbon reserves, and creating significant instability in the territory of the controlled area. The Arab Spring led to the disintegration of several states that were founded on traditional military-authoritarian statehood and the coming to power of radical Islamists, as well as a massive influx of refugees, mainly to the countries of Europe, which led to terrorist attacks in France (November 13, 2015, in Paris and July 14, 2016, in Nice), Belgium (March 22, 2016, in Brussels), Germany (December 19, 2016, in Berlin), and may lead to the fascisization of several European countries and the slow destruction of Europe.

In June 2012, a soft coup to change the president was carried out in Paraguay.

In 2013–2014, a soft (color, velvet) scenario of an information war was fully implemented in Ukraine with the aim of finally separating it from Russia.

In 2014, for the first time, China also began to implement a soft (color, velvet) scenario of an information war in Thailand and was successful.

Information wars continued in 2015 (Macedonia, Armenia, Moldova, [and] Venezuela) and 2016 (Kazakhstan, Turkey, Brazil, [and] again Armenia).

Of the 39 states in which wars were carried out and are continuing, in 28 wars, the main methods of influence (defeat) were methods and technologies (weapons) of informational-psychological influence (82 percent). Conventional methods of defeat were used in seven conducted and ongoing wars (18 percent) when it was not possible to achieve success by using methods and technologies (weapons) of informational-psychological influence and the usual methods of defeat began to be applied.

In 21 concluded wars and armed conflicts (54 percent) in the form of soft (color, velvet) revolutions (information wars) and power [revolutions], the goals were achieved. Of these, in 17 wars (81 percent), the goals were achieved through the use of methods and technologies of informational-psychological impact without the direct occupation of the territory of the state by the armed forces of the United States and NATO. At the same time, in power [revolutions] (using conventional weapons), the goals were achieved in relation to four states (19 percent) (Kuwait—1991, Yugoslavia—1999, Iraq—2003, [and] Libya—2011). The goals were not achieved in Georgia's war against South Ossetia. Conventional means of destruction have been used intensively in Syria since 2011 and since October 17, 2016, in Iraq during the liberation of the city of Mosul and other settled areas from ISIS [the Islamic State in Iraq and Syria], and informational-psychological means of influence have an auxiliary character.[2,3]

Based on the foregoing, the following conclusions can be drawn.

Today:

---

[2] V. K. Novikov, *Informatsionnoye oruzhiye - oruzhiye sovremennykh i budushchikh voyn [Informational Warfare: The Warfare of Contemporary and Future Wars]*, Moscow: Goriachaia Liniia—Telekom, 2011.

[3] V. K. Novikov, "Nie ubit, no podchinit [Not to Kill, but to Subjugate]," *Voenno-promyshlennyi courier [Military-Industrial Courier]*, December 16, 2015.

1. Information wars are becoming an objective reality of the present development of humanity, along with wars waged using conventional weapons. Until recently, it was hard to believe that information wars would be dominant in the achievement of the political, economic, territorial, and other goals of the United States and NATO. But this historical fact has already occurred. This is also facilitated by the fact that, at present, humanity is entering an era of a new technological paradigm—informational, which is leaving the embryonic phase [of development] and rapidly entering a phase of growth.
2. At the foundation of information wars is the condition for the creation of a revolutionary situation in the "victim" state, which will make it possible to peacefully change the government to the necessary one. At the same time, as a rule, the result of an information war is that the victim itself must thank the victor for the fact that it was defeated through appropriate contributions: sovereignty, mineral resources, etc.
3. Information warfare in modern conditions is becoming a permanent phenomenon, and it is viewed as an alternative to military activities. The purpose of these activities is not to destroy the enemy, but to reflexively control him in the interests of unconditional submission. Information wars are and will be fought with no less ferocity than wars that use conventional weapons.
4. The immediate unleashing of hostilities of any scale (if necessary) will be preceded by a set of nonmilitary measures (large-scale informational-psychological influence), which, according to its impact, will be comparable to the direct use of military force.
5. In modern conditions, the concepts of "operation," "battle," "systematic hostilities," and other military categories, in the classical sense, often become inapplicable in describing the processes taking place in the world and the resolution of emerging conflicts, especially [those that are] political, economic, territorial, national, [and] religious. These categories no longer fully meet the realities of today's world and are obstacles to the development of the forces and means of state armed forces, methods, and the forms of their application.

6. Preparation for a classical war—a war with the use of conventional weapons and nuclear deterrence—is no longer the full basis for maintaining the sovereignty of any state.
7. Information wars allow the United States and the West to enrich themselves at the expense of countries in which controlled chaos takes place, and thereby [the United States and the West] avoid crises or significantly reduce their consequences and restrain [their] competitors.
8. In current wars and armed conflicts, a tendency to transition to information wars has rapidly emerged, and it is necessary to draw the right conclusions from this.
9. The modern power and basis of the state, which enables the real, practical possibility of waging information wars without regard to the UN [United Nations] and other countries of the world, presupposes that the state has: (1) strategic nuclear forces (SNF); (2) forces and means of antimissile defense (ABM); (3) high-precision weapons (HPW); (4) naval forces (a Navy); (5) special purpose units (special operations forces); (6) reconnaissance by technical means; (7) developed special intelligence services; (8) navigation aids; (9) forces and means of EW; (10) global systems and means of communication (information and telecommunication systems); (11) powerful high-speed computers and integrated software; (12) informal, global, international media centers; (13) military bases abroad; (14) numerous human rights organizations (foundations) [and] human rights defenders abroad; (15) the industry of motion pictures, video, audio, and computer (virtual) games; (16) private military companies; (17) systematic training (education) of personnel in organizing and conducting information wars; (18) a developed theory of and real practice in waging information wars; (19) a legal framework that allows information wars to be waged in peacetime and wartime; (20) organizational structures (commands) for information warfare; [and] (21) great, world-class scientists, in particular, Nobel Prize winners.

At present, the United States is practically the only country in the world that, under present conditions, has the power and basis for waging information wars anywhere in the world to achieve its sociopolitical, economic, ideological, territorial, national, ethnic, religious, and other goals, which are founded in the global domination of the economy, financial, political, military, [and] technological sectors, [and] the dominance of American currency.

This power led the 44th U.S. President, Barack Obama, on September 10, 2013, when addressing the nation, to declare the exceptionalism of the American nation. The fact that the American nation is exceptional was also confirmed by the President of the United States during his speech on September 24, 2013, at the UN General Assembly. This rhetoric was continued by President Barack Obama in 2014–2016.

The 45th President of the United States, Donald Trump, in his inauguration speech on January 20, 2017, reaffirmed the U.S. policy of world domination. This confirms the sanctity of the U.S. policy for world domination, laid down in 1823 by the fifth U.S. President, J. Monroe. More than 190 years have passed—the policy is unchanged. The power of the United States allows it to divide all countries of the world into three categories:

- states loyal to the United States that fully support U.S. policy, especially in the international arena. These are the states of the British Commonwealth, Western Europe, Japan, Korea, Israel, Saudi Arabia, the [United] Arab Emirates, and several other states.
- "underdeveloped" states that need to be taught American values through political education and even coercion to accept these values. These are the states of Eastern Europe, Latin America, Africa, and Asia.
- states disloyal to the United States ("rogue states") that are not subject to American influence, do not want to adopt American values, strive to defend their national independence, and resist the dictates of the United States. These states are Russia, China, India, several states in North Africa, [and] the near and Middle East.

[This] power enables the United States to detain citizens of any state in various countries of the world at the request of law enforcement agencies, to extradite and sentence them to long terms, and also to extend the domestic laws of the United States to the entire world.

For example, just in 2013, Russians abroad were arrested and extradited to the United States—Dmitry Ustinov from Lithuania, Dmitry Belorossa from Spain, Maxim Chukharev from Costa Rica, Alexander Panin from the Dominican Republic, [and] Dmitry Smilyantsa from the Netherlands. The policy toward Russia continued in 2014–2016.

[This] power allows the United States to conduct covert operations on the territory of other states without their permission. For example, on May 2, 2011, in the Pakistani suburb of Abbottabad, during Operation Neptune Spear—permission for which was personally granted by President Barack Obama—Osama bin Laden, the leader of the Islamic terrorist organization Al-Qaeda, was killed. The operation was carried out without the consent and notification of Pakistani government authorities. On July 15, 2014, near Benghazi in Libya, special forces of the U.S. armed forces captured a citizen of that country, Ahmed Abu Khatil, and took him. Ahmed Abu Khatil was allegedly involved in the assassination of the U.S. Ambassador to Libya, Christopher Stevens.[4]

The power of the United States allows it to

- destroy all dissenting, hesitant, and doubting states that do not support or respect the United States
- have prisons (camps) in many countries of the world where people are held and tortured
- carry out strikes on the territory of independent states (September 23, 2014, in Syria), bypassing the UN
- present the negative consequences of their activities in any country in the world as the result of someone else's intrigues, but not those of the United States. What the United States does is all right and not subject to doubt
- define opponents and winners, appoint a state "victim," and determine their punishment.

Many countries of the world, taking into account the power of the United States—including in waging information wars—become allies of

[4] V. K. Novikov and S. V. Golubchikov, "Migrant istiny [Migrant of Truth]," *Voenno-promyshlennyi courier [Military-Industrial Courier]*, July 29, 2016.

the United States; practice neutrality; and strive to build similar or superior strategic nuclear forces, missile defense, high-precision weapons, Navy, intelligence, navigation, communications, electronic warfare, etc., [to those of] the United States.

China, Russia, and several other countries of the world, to preserve their sovereignty and integrity, as well as their ability to resist information wars or wage them independently or with allies, are taking vigorous, adequate measures to build up efforts in this sphere.

Prospects for tomorrow:

1. In the course of wars (military conflicts), there will most likely be no traditional, armed invasion in the form of strategic and other operations; tank, air, sea, and other battles; [or] deep inroads in groupings of troops and their surroundings.
2. At stake are not tanks and rifle complexes. Relying solely on nuclear missiles is no longer possible. The bet should be comprehensive.
3. Informational-psychological aspects of war cannot be regarded as an "add-on" for wars that are waged with the use of conventional weapons or rapidly growing cyber wars.
4. Information wars against states that possess nuclear weapons and their means of delivery are becoming a priority and most appropriate, because it is practically impossible to use nuclear weapons against numerous terrorists, insurgents, representatives of private military companies, other armed groups, [and] possibly many migrants, within the state.
5. Means and technologies used in the preparation and conduct of information wars will be systematically and efficiently developed and perfected; scenarios, forms, and methods of their application will fully rely on breakthrough technologies of the 21st century.
6. Training of personnel for conducting information wars in the long term will be the foundation of success.
7. Unfortunately, the Doctrine of Information Security of the Russian Federation, approved by decree of the President of the Russian Federation, No. 646, on December 5, 2016, does not fully reflect the issues of information wars and protection from them.

8. It would be desirable, to implement the provisions of the Doctrine of Information Security of the Russian Federation, for amendments and additions to be made to a number of regulatory acts of the Russian Federation: the FKZ [Federal Constitutional Law] from January 30, 2002, No. 1-FKZ, "On Martial Law;" the FKZ from May 30, 2001, No. 3-FKZ, "On Emergency Regulations;" the FZ [Federal Law] from December 29, 2010, No. 390-FZ, "On Security;" the FKZ from July 27, 2006, No. 149-FZ, "On Information, Information Technologies, and Defense of Information;" the ZRF [Law of the Russian Federation] from December 27, 1991, No. 2124-I, "On Mass Media;" "The National Security Strategy of the Russian Federation," approved by decree of the President of the Russian Federation, December 31, 2015, No. 683; and the "Military Doctrine of the Russian Federation," approved by the President of Russia on December 26, 2014, No. Pr-2976; etc., which would reflect the issues of regulating relations in the field of preparation and conduct of information wars.
9. To evaluate a secondary school according to a single criterion—the number of school children taking first place in international Olympiads in physics and mathematics; and higher education according to the criterion—the number of world-renowned scientists, in particular, Nobel Prize winners trained and living in Russia. This is because a state that is capable of training such students and scientists will always dominate the world in the long run, including in the conduct of information wars in defense of its national interests.
10. To increase the number of hours in mathematics and physics in all secondary schools by two hours per week, which will allow the Russian Federation to identify and prepare gifted children, which will allow Russia to occupy a leading position in the world in the field of developing software (software systems and applications) and information and telecommunication technologies and systems and ensure its power in the information sphere.

CHAPTER NINE

Information Confrontation: System of Terms and Definitions

> The field of future battles is, first and foremost, information.
>
> —*Nikolai Ogarkov*[1]

> To describe in the language of diplomacy the complex procedures of interaction between the negotiating parties, when it comes to the destruction of the most complex technological systems, it is necessary not only to overcome the language barrier, but also the adequacy of the conceptual framework related to the intended target, the responsibility for which lies with scientists and engineers—the creators of weapons.
>
> —*Yuriy Solomonov*[2]

At all stages of historical development,[3] information [warfare] has [been] waged in almost all wars.[4] With the formation of the information society, the globalization of information processes, and the democratization of society itself, and with the participation of more and more people in sociopolitical

[1] M. A. Gareev, "Voyna i sovremennoye mezhdunarodnoye protivoborstvo [War and the Modern International Confrontation]," *NVO*, No. 1, 1998.

[2] Y. S. Solomonov, "Strategicheskaya tsel' [Strategic Goal]," Moscow: OOO Publishing House A4, 2014.

[3] [Lata, Annenkov, and Moiseev, 2019.]

[4] N. L. Volkovskii, *Istoriya informatsionnykh voyn [History of Information Wars]*, Vol. 1, Saint Petersburg: OOO Polygon Publishing, 2003a; N. L. Volkovskii, *Istoriya informatsionnykh voyn [History of Information Wars]*, Vol. 2, Saint Petersburg: OOO Polygon Publishing, 2003b.

life, it became obvious that a new phenomenon has developed in the social sphere, the main substance of which is information struggle. This struggle is constantly being waged, both in peacetime and in wartime, between states (and not only between adversary states, but also between allied states while protecting their own interests in the course of fighting for zones of political influence, for markets, over disputed territory, for strengthening the defense sphere, etc.) and within each state in the struggle for power and money, for the ability to control large masses of people, for control over production, and for income from the sale of manufactured products.

The broadest view of this problem presupposes the enduring presence of information struggle in one form or another throughout the history of mankind, from the moment of the emergence of armed warfare. Even in ancient times, commanders and thinkers noted the importance of achieving victory without battles. Today, it is legitimate to assert that the 21st century will be the era of information, and that, all else being equal, the achievement of strategic advantages by the state will depend on its information capabilities. In the latter context, the modern actualization of this phenomenon, caused primarily by the introduction of modern information technologies and the development of the global information space, is yet another stage of the development of [information warfare].

The experience of waging wars and armed conflicts in recent years has forced military specialists to turn to a deep analysis of this phenomenon, and specifically to analyzing the results of the effects of information tools on the operation of combat systems and complexes, as well as the ability of the country's armed forces as a whole to conduct combat operations in modern conditions. In recent years, many publications have appeared on the problem of information confrontation (struggle) [IP(b)],[5] which has led

[5] V. M. Barynkin and M. A. Rodionov, *Ponyatiynyy apparat teorii informatsionnogo protivoborstva (Informatsionnoy bor'by) [Concepts of the Theory of Information Confrontation (Information Warfare)]*, Moscow: VAGSh, 1998; A. V. Manoilo, *Gosudarstvennaya informatsionnaya politika v osobykh usloviyakh: Monografiya [Federal Information Policies Under Special Conditions: Monograph]*, Moscow: MIFI, 2003; V. F. Prokof'yev, *Taynoye oruzhiye informatsionnoy voyny: Ataka na podsoznaniye [Secret Weapon of Information Wars: Attacking the Subconscious]*, 2nd ed., Moscow: Sinteg, 2003; I. N. Panarin, *Informatsionnaya voyna, PR i mirovaya politika [Information War, PR and Global Politics]*, Moscow: Hotline—Telecom, 2006; L. V. Vorontsova and D. B.

to the emergence in the military lexicon of such new terms and concepts as information warfare, information confrontation, "information war," information weapon, information resource, information space, information domain, information security, etc. Almost every author provides their own definition of one or another concept. This significantly complicates the development of a unified understanding of both the problem as a whole and its components because different interpretations of similar basic terms used in the information domain do not allow for correctly formulating a "tree" of goals and clearly establishing the stages and deadlines for completing given tasks. Therefore, there is a practical need for the formation of a single system of terms, concepts, and definitions in the information confrontation domain, which, on one hand, would encompass the terms, concepts, and definitions already established in this area, and on the other hand, would be internally consistent.

In several studies,[6] an approach to the formation of a system of basic terms, definitions, and concepts in the field of [information confrontation (struggle)] is proposed, the essence of which is reduced to the following provisions:

1. For basic terms, definitions, and concepts, it is necessary to use definitions that do not cause fundamental contradictions among the many researchers involved in the various aspects of [information confrontation (struggle)].
2. The definition of basic terms will need to be refined and concretized over time, taking into account the results of research in narrower

Frolov, *Istoriya i sovremennost' informatsionnogo protivoborstva [History and Modernity of Information Confrontation]*, Moscow: Hotline—Telecom, 2006; V. I. Annenkov, V. F. Moiseev, S. N. Baranov, and N. A. Sergeev, *Bezopasnost' i protivoborstvo v informatsionnoy sfere [Security and Confrontation in the Information Domain]*, Moscow: RUSAVIA, 2010; Novikov, 2011; V. I. Annenkov, S. N. Baranov, V. F. Moiseev, and S. S. Kharlakhoop, *Setetsentrizm: geopoliticheskiye i voyenno-politicheskiye aspekty sovremennosti* [Netcentrism: Geopolitical and Military-Political Aspects of Modernity], Moscow: RUSAVIA, 2013; and A. V. Manoilo, *Informatsionnyye voyny i psikhologicheskiye operatsii: Rukovodstvo k deystviyu [Information Wars and Psychological Operations: Call to Action]*, Moscow: Hotline—Telecom, 2018.

[6] Annenkov et al., 2010, p. 447; Annenkov et al., 2013.

areas of [information confrontation (struggle)]. Attempts to immediately obtain unambiguous definitions will introduce inevitable confusion, contradictions, and the unjustified creation of more and more terms that suit researchers.

3. The system of terms should evolve from general (sufficiently recognized) to specific (clarifying) terms, forming a hierarchical structure.
4. For the first basic concept in the field of [information confrontation (struggle)], one should take the concept of "information" (from the Latin *informatio*—explanation, presentation), which, without exaggeration, can be attributed to the most important systemological and philosophical categories because, at the present stage, it simultaneously includes three primary aspects: ordinary, natural-scientific, and philosophical.

In the ordinary sense, information is data about the surrounding world and the processes occurring in it, perceived by various consumers (humans, other living organisms, or special technical devices) to ensure purposeful activity.

In the natural-scientific sense, information can be defined as a property of matter, based on the fact that, as a result of the interaction of objects between their states, a certain relationship is established. The stronger this relationship, the more fully the state of one object reflects the state of another object, the more information one object contains about another. In this case, the physical carriers of information are signals, which are the states of physical fields or objects, and the relationship between the signal and the information it contains is established according to certain rules.

In the philosophical understanding, the concept of information has now acquired the meaning of an independent category. It is considered to be a fundamental property of matter, having the property of reflection.

Of the many definitions of the term information, let us focus on the following definitions.

Information (in a broad sense) is the property of objects (processes) of the surrounding physical world to generate a variety of states that are transmitted through reflection from one object to another (passive form) and

[are] a means of limiting both diversity and organization, i.e., management, disorganization, etc.[7]

Information (in the narrow sense) is information about persons, objects, facts, events, phenomena, and processes, regardless of the form of their presentation.

Review of the issues of preparation and conduct of combat activities shows that for [information confrontation (struggle)], a narrower definition of information is of interest.[8]

Specifically, information is data about objects and processes of any kind, including

- having a certain time-dependent utility for some decisionmaker (pragmatic aspect)
- reflecting essential (from the point of view of this [decisionmaker]) properties of objects or processes with a certain degree of accuracy and sufficiency (semantic aspect)
- presented with the help of a certain semiotic system (syntactic aspect)
- physically existing with the help of material-energy carriers (electromagnetic radiation, paper, magnetic carriers, etc.).

This clarification does not contradict the [aforementioned] definitions. It confirms them and makes it possible to highlight the [following] primary levels of information struggle:

- material-energy: struggle at the level of information carriers, that is, all types of hiding information and destruction of information systems, channels, and the information in them
- syntactic: struggle at the level of the structures of semiotic systems; that is, all types of codes, the use of ciphers, etc.

---

[7] Annenkov et al., 2010.

[8] Annenkov et al., 2010; [Government of the Russian Federation], "Federal Law No. 149, Ob informatsii, informatsionnykh tekhnologiyakh i o zashchite informatsii [On Information, Information Technologies and Information Protection]," Moscow, July 27, 2006.

- semantic: struggle at the level of the semantic content of information; that is, providing the enemy with meaningless or unreliable information (disinformation)
- pragmatic: warfare at the level of usefulness of information; that is, either changing the enemy's goals in relation to the use of information, or providing [the enemy] with useless information.

Information exists and moves in a certain part of the physical world, which is called the information environment. The geophysical concept of "environment" should not be confused with the concept of "domain," which defines the boundaries of the interaction of opposing sides in a specific "environment" (environments).

The movement of information in space and in time is manifested through the processes of search, collection, storage, processing, provision, accumulation, dissemination, and decisionmaking—which we will call information processes—and the space in which they appear is the "information space."

It is noted in [Manoilo's work] that in the informational aspect, the understanding of the term "information space" is based on the definition of the information domain.[9]

Let us define the information sphere as a part of the information environment in which information processes arise, develop, exist, and disappear, and, in turn, are the result of the relationship of specific sides (states, coalitions of states, ministries, departments, and other entities) in accomplishing specific tasks.

According to [the Information Security Doctrine of the Russian Federation], the information environment of the information domain includes the following elements (objects):[10]

- subjects of information interaction or influence (people, organizations, systems)
- information intended for use by subjects of the information sphere

[9] Manoilo, 2003.

[10] [Government of the Russian Federation], "Doktrina informatsionnoy bezopasnosti Rossiyskoy Federatsii [Information Security Doctrine of the Russian Federation]," No. 646, December 5, 2016.

- technical means, implementing information technologies
- information infrastructure that provides the ability to exchange information between subjects
- social relations that develop in connection with the formation, transmission, dissemination, and storage of information, and the system of their regulation.

Taking into account the [aforementioned], the concept of "information sphere" can be defined as a set of elements of the information environment (information, information infrastructure, subjects of information interaction [influence], technical means of implementing information technologies, and mechanisms for regulating relevant public relations) that are located on given territory and ensure the formation and existence of information processes, which are the consequence of the relationship between specific parties (states, coalitions of states, departments, agencies, and other entities) in solving specific problems.

From the [aforementioned], it follows that the concepts of "information domain of the Russian Federation" and "information space of the Russian Federation" are synonyms because their boundaries are uniquely defined.

To accomplish specific tasks, certain data are also required, concerning both friendly and enemy troops and information about the characteristics of the environment in which the opposing sides interact. For convenience, these data (information) are collected in databases and appropriately (with the help of technical means that implement the relevant information processes) provide access to them for governing bodies (decisionmakers), i.e., creates an information system.

The set of information systems—united according to a certain rule in a supersystem for obtaining information that is structured by certain rules in the information domain and intended for governing bodies (decisionmakers) to accomplish specific tasks—will be called the information field.

The Military Doctrine of the Russian Federation sets out the task of "qualitatively improving the means of information exchange in a single information space of the Armed Forces, other troops and bodies as part of the information space of the Russian Federation."[11]

Let us clarify the content and correlation of the concepts "information domain of the [Russian Federation] Armed Forces," "information space of the [Russian Federation] Armed Forces," and "single information space of the [Russian Federation] Armed Forces." We have shown [previously] that the concepts "information domain of the [Russian Federation] Armed Forces" and "information space of the [Russian Federation] Armed Forces" can be considered synonyms. But the concept of "single information space of the [Russian Federation] Armed Forces" has its own characteristics (features) that ensure the interaction of military command and control bodies at various levels in accomplishing a wide range of tasks under various conditions.

Taking into account the [aforementioned] approach to the formation of a system of terms, as well as the content of terms and definitions used in [the Military Doctrine of the Russian Federation], the substance of this concept can be defined as follows: A single information space of the [Russian Federation] armed forces is the information space of the [Russian Federation] armed forces that is part of the [Russian Federation] information space and represents a set of information fields, each of which is created and functions by common principles and general rules, ensuring the communication of military command and control bodies at various levels (both vertically and horizontally) regarding the challenges that they face under various conditions and in accordance with valid normative legal documents.[12]

The term "cyberspace" is widely used when emphasizing the interaction between opposing parties in the information space. It was originally used in science fiction to describe direct communication between the brain and the computer. In the mid-1990s, because of the global informatization of society (the development of the global computer network), a fundamentally new environment for the interaction of opposing parties emerged, called "cyberspace."

---

[11] [Government of the Russian Federation,] "Voyennaya doktrina Rossiyskoy Federatsii [Military Doctrine of the Russian Federation]," No. Pr-2976, December 25, 2014.

[12] Government of the Russian Federation, 2014.

Cyberspace cannot be recognized as a kind of real physical space. It exists only in our minds, claiming to be a reflection of real space as a kind of virtual space. The word “virtual” (from the Latin *virtus*—strength, ability) means “able to be,” but in reality has no place, existing only potentially.

Successful practical experience suggests that cybernetic (virtual) space adequately reflects real space. It is an element of the information environment and is a special kind of environment; it differs from other environments (sea, land, air, space) with its artificial, man-made creation and requires continuous efforts to maintain itself.

Cyberspace as an environment interacts with other environments and is a kind of “connecting environment” for all other environments, ensuring the formation of a single image of the situation because such an image is directly formed and maintained in cyberspace. Cyberspace does not have an extension but it is directly related to physical space because its infrastructure (for example, servers) is located in this space and it also reflects the presentation of physical space.

From the [aforementioned], it follows that cyberspace is a kind of virtual space that adequately reflects the elements of a real, artificially created information environment as a set of information structures that include telecommunication networks, computing systems, processors, and controllers embedded in hardware, [and] plays the role of a “connecting environment” for all other environments (sea, land, air, space), ensuring the formation of a single image of the situation.

The terms “information confrontation,” “information warfare,” “information war,” [and] “network-centric war” are widely used in mass media and in normative legal documents. In some cases, their use is associated with current trends, and in others, with an increase in the role and significance of the information component in relations between opposing sides (states). However, authors interpret the meaning of these terms differently when using them.

The term “information war” appeared in the mid-1980s in connection with new challenges for the U.S. armed forces, came into active use after Operation “Desert Storm” in 1991, and was officially codified in U.S. Department of Defense directives in December 21, 1992, and in [Department of Defense Directive] TS 3600.1 as a set of activities designed to achieve information superiority in support of the national military strategy by influencing the enemy's information and information systems and, at the same time, ensuring the security and protection of one's own information and information systems.

[When] defining the concept of "information war" and justifying its use in describing a new phenomenon, we should remember that the key word in this concept is "war." And war is a sociopolitical phenomenon, a special state of society associated with a sharp change in relations between states with the transition to the use of armed force to achieve political goals. The state of war begins with initiation of hostilities and ends when they conclude. According to Article 18 of the [Federal] Law [of the Russian Federation] "On Defense," "a state of war is declared by federal law in the event of an armed attack on the Russian Federation, another state or group of states, as well as, if necessary, the implementation of international treaties of the Russian Federation." Along with the armed warfare that makes up the core content of war, economic, diplomatic, ideological, information and other "nonmilitary" means can be used prior to or during war and can take on a fiercer character in wartime.[13]

The phenomenon we are considering has its own specifications (features):

- Activities within the framework of this phenomenon are carried out constantly (in peacetime, in crisis situations, during the preparation and conduct of war), in various areas of life, as a rule, by specialized (both military and civilian) structures and using specific activities and special resources.
- The main goal is to achieve information superiority by influencing information and information systems of the opposing side (enemy) while simultaneously protecting one's own information and information systems.
- Objects of the information domain can act both as offensive and defensive entities.
- [The phenomenon involves] concealing the influence on the opposing side and the ability to use special means on a mass scale.
- [It also involves] the capability of delivering direct impact on the human psyche.

[13] Government of the Russian Federation, 2014.

The scale of application, decisiveness of goals, versatility (the ability to apply to solve a wide range of problems), effectiveness of impact, and use in various spheres of human activity allowed some politicians and political scientists to talk about this phenomenon as "information war." But this is not a war in the classical sense of the word because its primary component is missing—armed warfare. It is most correct to speak of this phenomenon as information confrontation (information warfare in military affairs). The President of the Academy of Military Sciences, General of the Army, M. A. Gareev has spoken about this topic. In [Gareev's writings], he notes, "Given the increasing scale and effectiveness of information . . . and other non-military means of influence in international confrontation . . . some scientists and political figures are raising the question of revisiting the essence of war and some fundamentals of the study of war. . . [they are] propos[ing] to revise the definition of the very concept of war, assuming that any international confrontation, including . . . using information means, is war. . . . [They suggest], for example, that any cybernetic intrusion into the internet and other information actions be considered equivalent to a declaration of war, and [if] such actions are occurring constantly, then all countries will find themselves in a state of permanent war with each other."[14]

The definitions of the terms "information confrontation" and "information warfare" should follow from the need to take into account the following components (elements) of this phenomenon:

- This is, on the one hand, a characteristic of the relationship (confrontations) between the opposing sides in the information domain and, at the same time, a set of activities (measures) aimed at resolving them.
- On the other hand, it is a set (complex) of measures carried out by certain forces in the information domain using special means.

[14] M. A. Gareev, "Itogi deyatel'nosti Akademii voyennykh nauk za 2012 god i zadachi Akademii na 2013 god [Activities of the Academy of Military Sciences for 2012 and Academy Tasks for 2013]," *Vestnik Akademii voennykh nauk [Journal of the Academy of Military Sciences]*, Vol. 1, No. 42, 2013a; M. A. Gareev, "Sistema znaniy o voyne i oborone strany na sovremennom etape [System of Knowledge about War and Defense of the Country in Modern Times]," *Vestnik Akademii voennykh nauk [Journal of the Academy of Military Sciences]*, Vol. 2, No. 43, 2013.

- The measures correspond to a single concept and plan, are aimed at achieving a specific goal (achieving information superiority in command and control), and are carried out continuously, both in peacetime and in wartime.
- Measures carried out within the framework of the [Russian Federation] armed forces are coordinated with public authorities.
- The measures deliver a specific impact against enemy targets or the protection of their own targets and, ultimately, ensure the achievement of the objective, i.e., achieving information superiority.

Then the concept of "information confrontation" can be defined as the state of relations between the opposing sides in the information domain, primarily consisting of specific impacts on the information targets of each side and measures to protect them from the specific influences of the opposing side to achieve information superiority while solving problems to realize their interests.

We define "information warfare" as a set of activities and measures carried out by troops according to a single concept and plan to gain (maintain) information superiority through specific impacts on the enemy's information targets and protecting one's own information targets from the enemy's specific influences.

Based on the [aforementioned] and in the opinion of the authors, it follows [that]:

- It is advisable to exclude the term "information war" from use in official legal documents because this is not war in the classical sense of the word; its primary component, armed warfare, is missing.
- The term "information confrontation" is advisable to use when addressing the relationship between states in the information domain when they strive to achieve information superiority while accomplishing tasks [in accordance with] their interests.
- It is advisable to use the term "information warfare" in relation to the armed forces because it emphasizes the specifics of information confrontation in the preparation and conduct of military (combat) activities.

In the late 1990s, the term "network-centric war" appeared in our literature as a result of an inaccurate translation of the English term "centric network–warfare." A more accurate translation corresponds to the concept of "network-centric support of military operations." Consequently, we are not talking about some new type of war, but about a network-centric approach (principle) to the preparation and conduct of military operations. According to this approach, commanders, military units, [and] each tank and soldier should be united within the framework of battle (operation) by a single information network. This allows for a quick exchange of information [and allows the receipt] of all the necessary data about the enemy. The application of this approach makes it possible to increase the combat readiness and effectiveness of all military units.

Inaccurate translation also results in inaccurate understanding or even misunderstanding of the essence of the problem. Nevertheless, the term "network-centric war" has already taken root and the task now is not to change it but to understand and use it correctly. The concept of "network-centric warfare" is synonymous with the concept of "network-centric support to military operations."

Network-centric support to military operations is a set of technical and organizational measures interconnected in purpose, tasks, place, and time, and carried out among troops in peacetime and in wartime to combine command and communication networks, reconnaissance networks, and weapons into a single military-technical system using a given information field.

Network-centric support to military operations significantly increases the combat capabilities of troops in the conduct of military (combat) operations because of [its] ability to inform geographically dispersed commanders (decisionmakers) across the theater of combat operation[s] in a timely manner at various levels and in control of different weapon systems.

The target of influence in the framework of information confrontation (warfare) is the information resource. Analysis shows that various sources interpret the definition of this term differently, but basically everything comes down to the volume of documented information (documents); that is, to information already received, validated, and recorded in a physical medium.[15]

The extension of the definition of [an information resource] to the military domain unacceptably narrows the meaning of this concept and seems to be impractical. Indeed, the most significant amount of documented information is a small part of the volume of information available in managing organizations and various technical systems. Thus, orders and commands (signals) transmitted over communication lines, such as information from space reconnaissance and navigation and meteorological support systems, do not fit into the concept of the document. In addition, this definition does not take into account such important aspects of [information resources] as the capabilities of generating, compartmentalizing, and reproducing information; the technical systems themselves that support information processes; and the personnel and decisionmakers using these systems.

Let us clarify the definition of the concept of "information resource" in relation to the military domain. The word "resource" carries the main semantic meaning in this concept. The dictionary indicates that this term has French origin (from the French *ressurce,* meaning "auxiliary means"), i.e., everything that is used with a purpose, [which] can include all purposeful activities of a person or people and the activity itself.[16] In other words, it is "a set of means that allow [one] to obtain the desired result with the help of certain adaptations." That is, an information resource is a set of components in the information domain that ensures the formation and flow of information processes, i.e., create information capabilities that, in turn, determine and support [both] the state (the actor in the given interaction) in achieving strategic advantages over other actors in defending its own interests and the armed forces in conducting military (combat) operations under modern conditions.

---

[15] Barynkin and Rodionov, 1998; Manoilo, 2003; Prokof'yev, 2003; Panarin, 2006; and *Slovar' terminov i opredeleniy v oblasti informatsionnoy bezopasnosti [Dictionary of Terms and Definitions in the Area of Information Security]*, 2nd ed., Moscow: VAGSh, 2008.

[16] *Noveyshiy slovar' inostrannykh slov i vyrazheniy [New Dictionary of Foreign Words and Expressions]*, Minsk: Contemporary Literary Man, 2006.

Then the definition of the concept of "information resource" in relation to the military domain should include information systems (including a human operator and decisionmakers), information channels (formed by information transmitters, information receivers, and the information distribution environment) and the information itself that exists and circulates through information systems (libraries, archives, foundations, [and] databases) that make up the components of an information resource. We will call the components of an information resource information targets.

That is, an information target is a component of an information resource (organizational and technical systems, special devices, a decisionmaker, etc.) that ensures the flow of information processes, including the decisionmaking process. Depending on [their] composition, information targets are subdivided into informational-technical and informational-psychological. Informational-technical targets consist of information targets that include man-made informational-technical systems and means, and informational-psychological targets consist of information targets related to human activities, i.e., governing bodies of various levels, decisionmakers, operators and service personnel, and other actors.

This interpretation of [the concept of information resources] allows for a clearer definition of the goals of [information confrontation (warfare)], [which include] the selection of the adversary's targets and the means to deliver effects against them to disrupt their operation and ensuring information security (protection) of the information of one's own [information resource] targets in the interests of gaining (maintaining) information superiority. Effects against information targets ensure the destruction of the enemy's [information resources]. This solves the problem of not only disrupting information processes, but also delivering effects against the information itself (its reliability, completeness, integrity, timeliness, etc.). Protection of information targets from specific effects ensures the operation of [information resources] as a whole. We will call the specific effects that are delivered in the course of [information confrontation (warfare)] against a specific information resource (information targets) information effects.

Information effects are effects against [information resources] (information targets), the purpose of which is to disrupt information processes, including the decisionmaking process.

Information effects are created and delivered with the help of special means or tools that, in some sources, are called "information weapons," and in others "means of information confrontation (warfare)." Moreover, these concepts, especially the concept of the information weapon, are interpreted in different ways, for example, as:

1. A set of means used to disrupt (copy, distort, or destroy) information resources as they are being created, processed, distributed, and/or stored. The targets of its effects are software and information support; software and hardware, telecommunication and other means of informatization and command and control; communication channels ensuring information flows and the integration of command and control systems; [and] human intelligence and mass consciousness.
2. Means for destroying, distorting, or stealing information volumes, extracting the necessary information from them after overcoming protection systems, restricting or prohibiting access to them for legitimate users, disrupting the operation of technical means, disabling telecommunication networks, computer networks, [and the] entire high-tech system that sustains life in society and the operation of the state.
3. A set of information technologies, methods and means of information effects, intended for waging information war.
4. Methods and means of delivering informational effects against equipment and personnel to accomplish specific tasks.

Each of these definitions does not fully reflect the substance of the given concept since:

- The method cannot be viewed as the tool to influence equipment and personnel (definition 3 and 4)
- Definition 3 uses the concept "information war," which is itself incorrect
- Information technologies in their pure form do not fall under the concept of "information weapon," because Federal Law defines them as "processes, information research, collection, storage, processing, pre-

sentation, and dissemination methods and means to operationalize these processes and methods"[17]

- All of the concepts that are presented do not fully reflect the purpose of the [information] weapon. This is a means to achieve and maintain information superiority in the information domain, and not just to disrupt (destroy, distort, or steal) information resources; that is, it completely excludes the protective function.

Analysis of the definitions of well-known and tested concepts of various types of weapons made it possible to formulate the concept of "information weapon" as a set of technical, software, and other special means, constructively designed to create information effects to disrupt information processes by delivering effects against an information resource (information targets) through specially planned radiation emissions of various types of energy or in specially selected and structured information.[18]

At present, instead of the concept of information weapon, it is advisable to use the concept of "means of information effects," which is much broader than the definition of the concept of information weapons. This is because of the fact that the creation of information effects is not limited to means specially created for this.

Then the means of information effects should represent any means (technical device, linguistic and software products, medication, etc.) by which it is possible to create and deliver informational effects against the [information resources] (components of the information resource) of the opposing side or to protect one's own information resources (components of the information resource) to achieve (maintain) information superiority in the information domain.

Given the above discussion, the means of information effects include

- means of electronic warfare, including means of technical disinformation

---

[17] Government of the Russian Federation, 2006.

[18] *Sovetskaya Voyennaya Entsiklopediya [Soviet Military Encyclopedia]*, Vol. 8, Moscow: Voenizdat, 1980; Government of the Russian Federation, "Federal Law No. 150, "Ob oruzhii [On Weapons]," December 13, 1996.

- special software and other effects against automated control systems and digital technology
- psychotropic generators
- special pharmaceutical effects against the adversary's population
- mass information means (including audio and video data synthesizers), [including the] creation of holographic images in the atmosphere.

Information warfare is waged to achieve information superiority in the information domain, including in command and control. Information superiority in command and control is superiority in the following: timeliness, reliability, and full receipt of information by all command and control bodies; speed and quality of information processing and timeliness of decisionmaking; timeliness and delivery of decisions (orders) to those who will execute them and the reliable monitoring of their execution.

The following ensure the achievement of information superiority:

- developing and maintaining information resources in the state that allow [the state] to make appropriate decisions, as well as to ensure the moral and psychological resilience of decisionmakers and the entire population in the interests of achieving political aims
- neutralizing (preempting or decreasing) information effects delivered by the adversary against information resources
- coordinating effects with departments and agencies against adversary information resources, [thereby] forcing adversary political leadership to make political, military, technical, economic, and other decisions [that are] favorable to the Russian Federation.

The concept of "information security" is used widely to determine the level of protection of information resources (information targets) from information effects. It has become the most important component of national security of all leading world countries, including the [Russian Federation].

Let us clarify the semantic definition of this term. The key word here is security. Security is the state of the protected object (information target) when harmful effects targeted against it (energy and information flows) do not exceed the maximum allowed levels, and the information resource as a whole and its individual components comprise the protected objects. In that case, information security is the state of the protected object (information target) when harmful effects targeted against it (information effects) do not disrupt its normal operation (separate components of the information target).

A harmful effect is a negative effect against an individual (operator or decisionmaker) that leads to degradation of health, illness, or death of the individual (operator), as well as against a technical system, which decreases [the system's] operational capabilities or destroys it.

The prior discussion asserts that the information security of the state (state's armed forces) [consists of] the protection of its information resources (information targets) from information effects and from planned and accidental individual actions, focusing on collection, processing, storage, and delivery of information, as well as on decisionmaking in any situation (in wartime and in peacetime).

Information security is assured either through eliminating information threats and, if [there are] any, providing sufficient protection against them.

CHAPTER TEN

Analysis of the Organization and Conduct of Informational-Psychological Operations in the Conduct of Hybrid Wars

Introduction

The combination of conventional means of warfare with informational and psychological actions and operations leads to the hybridization of warfare. This phenomenon is conceived of as "hybrid war."[1] The current hybrid war of Russia against Ukraine, disguised as a popular uprising of the so-called Donetsk People's Republic and Luhansk People's Republic, is a clear example of such. It demonstrates a new approach to the conduct of military campaigns with psychological and informational "cultivation" of the local population as a key component. It is aimed at the destruction of the values and suppression of the will of the people.[2]

---

[1] [I. Yuzova, 2020]; Volodymyr Horbulin, *Svitova hibrydna viina: Ukrainskyi Front [The World Hybrid War: Ukrainian Forefront]*, Kyiv, Ukraine: National Institute for Strategic Studies, 2017, p. 496; Fedir Turchenko and Halyna Turchenko, *Proiekt "Novorosiia" and novitnia Rossiisko-Ukrainska viyna [The Novorossiya Project and the Latest Russian-Ukrainian War]*, Kyiv: Institute of History of Ukraine, 2015.

[2] Horbulin, 2017; Turchenko and Turchenko, 2015; D. Prysiazhniuk, "Zastosuvannya manipuliatyvnykh psykhotekhnolohii z boku Rosii v ZMI Ukrainy (Na prykladi Krymu) [Application of Manipulative Psychotechnologies by Russia in Ukrainian Media (on the Example of Crimea)]," *Visnyk Kyyivs'koho Natsional'noho Universytetu*

Therefore, the Russian military and political leadership gained an advantage during the Crimea annexation and occupation of part of the Donetsk and Luhansk regions with nearly the same level of traditional armaments and military equipment because of informational-psychological operations (henceforth—IPO) on the national (strategic) and operational and tactical levels;[3] [these actions were] planned and carried out in advance. Therefore, modern hybrid means of warfare gain even more significance during contemporary conflicts.

It is necessary to have a clear understanding of the nature and features of this conflict to resist, deter, and protect both the Ukrainian population at large and armed forces personnel from negative informational-psychological influence (henceforth—IPI).

Analysis of the Latest Research and Publications

The number of scholarly publications on the topic of hybrid warfare increased sharply after 2014 and continues to grow,[4] but the specifics of the organization and conduct of the hybrid war against Ukraine remain unanswered. Having a large number of facts and data about hybrid military actions, scholars face difficulties in their interpretation. As a result of the actions of the Russian Federation against Ukraine, some definitions of the nature of hybrid warfare need to be refined. First of all, [the term] applies to the asymmetric nature of hybrid threats by a weaker enemy against a party

---

imeni Tarasa Shevchenka. Viys'kovo-spetsial'ni nauky [Journal of Taras Shevchenko National University of Kyiv. Military Sciences], No. 23, 2009.

[3] H. Pievtsov et al., *Informatsiino-psykholohichna borotba u voiennii sferi [Information and Psychological Warfare in the Military Domain]*, Kharkiv: Kharkiv National University, 2017; H. Yavorska, "Hibrydna viina yak dyskursyvnyi konstrukt [Hybrid Warfare as a Discursive Construct]," *Stratehichni priorytet [Strategic Priorities]*, No. 4, 2016.

[4] Horbulin, 2017; Turchenko and Turchenko, 2015; Prysiazhniuk, 2009; Pievtsov et al., 2017; Yavorska, 2016; V. Stasiuk, *Psykholohichne zabezpechennia diialnosti viisk (syl)" [Psychological Support for the Activities of Troops (Forces)]*, Kyiv: Ivan Chernyakhovsky National University of Defense of Ukraine, 2014.

with more-powerful military forces, the latest technologies, and high demographic potential.[5]

A significant contribution to the development of the theoretical foundations of hybrid conflicts has been made by domestic [Ukrainian] researchers.[6] Most of them continue their research, using the experience of the antiterrorism operation [official name of military operations in Eastern Ukraine].[7]

The purpose of this article is to study the process of organizing and conducting informational-psychological operations as a part of hybrid warfare.

Presentation of the Main Material

Hence, *hybrid warfare* should be defined as military actions that combine military, diplomatic, information, economic, and other means to achieve strategic policy goals. The uniqueness of this combination lies in the fact that each of the military and nonmilitary means of hybrid warfare is used as a weapon.

The use of the armed forces to achieve certain objectives by military means is a historical phenomenon. [At] the current stage of scientific and

---

[5] M. Dziuba et al., *Narys teorii i praktyky informatsiino-psykholohichnykh operatsii [Essay on the Theory and Practice of Information and Psychological Operations]*, Kyiv: VITI NTUU "KPI," 2006; M. Libiki, "Shcho take informatsiina viina? [What Is Information Warfare?]," 2019; Oleksandr Lytvynenko, "Totalna viina po-Putinski: 'Hibrydna' viina RF proty Ukrainy [Total Putin-Style War: Russia's 'Hybrid' War Against Ukraine]," in *"Hibrydna" viina Rosiii—vyklyk i zahroza dlya Yevropy [Russia's Hybrid War—Challenge and Threat for Europe]*, Kyiv: Razumkov Centre, 2016; A. Barovska, ed., "Informatsiini vyklyky hibrydnoi viiny: Kontent, kanaly, mekhanizmy protydii' [Information Challenges of Hybrid Warfare: Content, Channels, Counteraction Mechanisms]," Kyiv: National Institute for Strategic Studies, 2016.

[6] Dziuba, 2006; Pavlo Hai-Nyzhnyk et al., *Ahresiia Rosii proty Ukrainy: Istorychni peredumovy ta suchasni vyklyky [Russia's Aggression Against Ukraine: Historical Background and Current Challenges]*, Kyiv: MP Lesya, 2016.

[7] Turchenko and Turchenko, 2015; Prysiazhniuk, 2009; Pavlo Hai-Nyzhnyk, *Rosiia proty Ukrainy (1990–2016 r.): Vid Polityky shantazhu i prymusu do viiny na pohlynannia ta sproby znyshchennia [Russia Against Ukraine (1990–2016): From a Policy of Blackmail and Coercion to a War for Absorption and Attempts to Destroy]*, Kyiv: MP Lesia, 2017.

technical development, the informatization of all aspects of society has given rise to a genuine revolution in the military domain. The concept of a total war exhausted itself historically because further large-scale use of weapons against armies and populations in modern wars leads to catastrophe on a global scale [and] the death of civilization and the environment.[8] The world is entering a new phase of next-generation wars, aimed not so much at direct annihilation of the enemy but instead at the achievement of political objectives of war without the use of mass armies.

To achieve established political objectives in war, various means are used—economic, diplomatic, psychological, and other—and corresponding forms of combat to reduce costs, [as well as] the use of one's own forces and means. And the resolution and achievement of one's own geostrategic goals, security, sovereignty, and territorial integrity is done, for the most part, through battles on the information field.

The conduct of IPOs becomes an integral part of all military actions. Under current circumstances, the conduct of IPOs prior to armed conflict has become a convention. The leading countries of the world employ weapons only after a preparatory information campaign, when the information domain of the adversary is significantly deteriorated and the risk of defeat is fully eliminated. In this case, IPOs play a decisive role.

Informational-psychological operations, as defined by Balabin,[9] should be treated as a system of information acts, attacks, and actions that are internally coherent and interconnected by purpose, tasks, objects, and time, and are conducted simultaneously or sequentially under a unified plan and for a joint purpose of IPI on the target audience. It should be noted that IPOs are conducted in peacetime and during a special period, i.e., long

---

[8] N. Voloshyna and M. Dziuba, "Vyroblennia u maibutnikh ofitseriv imunitetu proty nehatyvnoho informatsiino-psykholohichnoho volyvu [Development of Immunity of Future Officers to Negative Information and Psychological Influence]," *Visnyk Kyyivs'koho Natsional'noho Universytetu imeni Tarasa Shevchenka. Viys'kovo-spetsial'ni nauky [Journal of Taras Shevchenko National University of Kyiv]*, No. 30, 2013.

[9] Y. Zharkov et al., eds., *Istoriia informatsiino-psykholohichnoho protyborstva [History of Information and Psychological Confrontation]*, Kyiv, 2012.

before the open military combat. Using an analysis of known sources of information,[10] [we present] the main objectives of IPOs.

In reality, Korea (1950–1953), Afghanistan (1979–1989), Grenada (1982), Panama (1989), the Persian Gulf "Operation Desert Storm" (1991–1992), Yugoslavia "Operation Allied Force" (1999), Iraq "Operation Iraqi Freedom" (2003), and Afghanistan "Operation Enduring Freedom" (2001–2002), as well as Ukraine's IPOs, [were] executed to create conducive conditions for further successful operations and combat; [to show] effective use of one's own troops (forces), armaments, and military equipment; and [to reduce] the effectiveness of the enemy's use of troops (forces) and weapons by achieving and retaining an information advantage over the adversary during training and military (combat) operations through the indirect pull of the adversary under one's own information control.

To conduct IPI against an adversary during IPO, various means are used, namely

- mass media
- internet sources
- printed materials
- various audio and video products
- specially trained and selected people who conduct information and propaganda influence by personally communicating with people.

At the same time, special consideration is given to the most influential individuals, political leaders, journalists, and, especially, military personnel of all ranks.

The organization and conduct of IPOs includes the following features:

- unexpectedness
- latency
- absence of aggression
- impossibility of bringing the aggressor to justice
- serious damage to the object of influence without the declaration of war or breaking diplomatic relations

10 Dziuba et al., 2006.

- significant difficulties in identifying the sources of aggression, its scale and goals, methods and forms, instruments, and means of conducting IPOs.

Currently, IPOs have become an integral part of the art of war. They are aimed not at the physical annihilation of the adversary and not even at destruction of [the adversary's] important strategic and economic objects, but, above all, at laying the groundwork for further implementation of one's own military campaign using armed forces. Informational-psychological warfare has become the central meaning of war [because] it is an integral part of a full-scale war and puts information weapons on par with weapons of mass destruction. Therefore, protection against the negative IPI [influence] of IPOs is of paramount importance and requires the state to use all possible measures to provide informational-psychological security to mili-

FIGURE 10.1
Main Objectives of IPOs

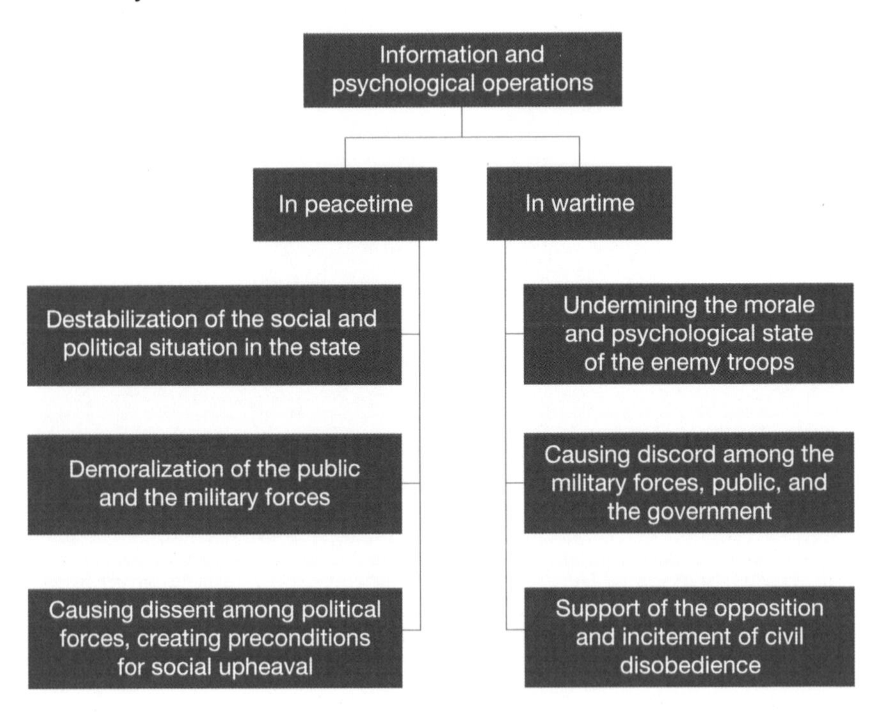

tary personnel in Ukraine and society as a whole as a component of the information security of the country.

The information technologies that revolutionized the communication space not only provided great opportunities for information exchange but also created the environment for effective IPI.

Informational-psychological influence forms the essence of psychological operations because it is a mode of implementation. Informational-psychological influence is such an influence on individual or social consciousness that [it] transforms the psyche [and] changes views, opinions, relationships, values, motives, attitudes, [and] stereotypes of the object to affect [its] actions and behavior.

One of the distinguishing characteristics that has developed under modern conditions—not only in Ukraine but also in the world writ large—is an advanced development of forms, means, technologies, and methods of influencing consciousness (subconsciousness) and the psychological state of an individual in counteraction to the organization of resistance to negative psychological influences, informational-psychological protection of a military employee and society as a whole. The need to protect the personnel of the armed forces from the negative IPI of the enemy is especially important. This need is caused by the active development and implementation of various forms, methods, means, and techniques of IPI on individual, group, and mass consciousness that can be observed in the military domain. The Russia-Ukraine conflict, which started in 2014 with the annexation of Crimea and continues today in east Ukraine, perfectly illustrates the entire arsenal of contemporary "hybrid wars."

This influence is a set of specifically planned and prepared acts, actions, [and] operations (measures, forms, and methods) of influence [that target] civil and military information infrastructure, individual or social consciousness, the moral and psychological state of troops (forces), and the population of the adversary to achieve military, political, economic and psychological goals.

It is the change of the system of mass information and the dissemination of false, untruthful information that is misinforming the society, which leads to the disruption of social stability [and] harm to the health and life of citizens as a result of propaganda or campaigning that incites social, racial, national, or religious hatred and enmity. These influences can lead and actually do lead to serious mental and physical health disorders, changes of generally accepted norms of behavior, [and] an increase in the number of risky social and personal situations.

Treating IPI as a threat, we first of all imply the negative consequences of its implementation. Negative consequences can manifest themselves in two aspects: first, in the attitude of the individual toward the government, and second, through the destruction of the integrity of the individual.

The ability to effectively influence the intellectual potential of the country, disseminate and instill into public consciousness the corresponding spiritual and ideological values, [and] transform and undermine the traditional foundations of nations and peoples are the ultimate goals of IPI, which is used for the purposes of IPOs.

Conclusion

Consequently, we can conclude that today, IPOs have become an integral part of the art of war. They are aimed not at the physical annihilation of the enemy and not even at destruction of its important strategic and economic facilities, but, above all, at creating the foundation for further implementation of one's own military campaign—but with the use of armed forces.

The application of modern means of information warfare is aimed at the distortion of the ideological sphere, spiritual and material values, change of societal and political systems, the collapse of the state and the army, the depletion of the economy, the destruction of the education system, [and] the exacerbation of ethnic and religious conflicts. That is why protection from negative IPI is of paramount importance and requires the government to use all possible measures to ensure the informational-psychological security of the individual and society as a component of the information security of the country.

The hybrid war that is being waged today against Ukraine undermines political, economic, and social stability in the country and leads to numerous casualties.

The reality of what is taking place in Ukraine today—the annexation of Crimea; fighting in Donbass; and [separatist forces'] colossal informational, psychological, and military support from the Kremlin—clearly demonstrates the [need to protect] personnel from the negative informational-psychological influence of the enemy.

Therefore, the organization and execution of IPOs has become a mandatory element of the moral-psychological support of the armed forces of Ukraine in conflicts of varying intensity, [including] peacekeeping [and] antiterrorist operations.

The direction of further research is to identify socio-psychological features of negative informational-psychological influence on all military units during combat.

Abbreviations

CW	conventional warfare
DDoS	distributed denial of service
DRG	sabotage and reconnaissance forces [diversionno-rezvelyvatel'noi gruppy]
EW	electronic warfare
FKZ	Federal Constitutional Law
IPI	informational-psychological influence
IPMT	informational-technological means and technologies of influence on an individual, society, [and] population
IPO	informational-psychological operations
NATO	North Atlantic Treaty Organization
RSCS	reconnaissance and strike combat systems
SSR	Soviet Socialist Republic
UN	United Nations
USSR	Union of Soviet Socialist Republics

References

Akhmadullin, V., "Kiberprostranstvo pod pritselom Pentagona [Cyberspace at Pentagon's Gunpoint]," *Nezavisimoe voennoe obozrenie [Independent Military Review]*, No. 1, January 12, 2007.

Akulinchev, A., "Problemy tsifrovizatsii voyennykh setey svyazi i puti ikh resheniya [Problems of Digitalization of Military Communication Networks and Ways to Solve Them]," *Voennaya mysl' [Military Thought]*, No. 9, September 2006, pp. 76–80.

Andreev, A. F., and I. S. Belobragin, "Informatsionnoye protivoborstvo i bezopasnost' gosudarstva [Information Confrontation and State Security]," *Vestnik Akademii voennykh nauk [Journal of the Academy of Military Sciences]*, Vol. 4, No. 17, 2006, pp. 21–28.

Annenkov, V. I., S. N. Baranov, V. F. Moiseev, and S. S. Kharlakhoop, *Setetsentrizm: geopoliticheskiye i voyenno-politicheskiye aspekty sovremennosti* [*Netcentrism: Geopolitical and Military-Political Aspects of Modernity*], Moscow: RUSAVIA, 2013.

Annenkov, V. I., V. F. Moiseev, S. N. Baranov, and N. A. Sergeev, *Bezopasnost' i protivoborstvo v informatsionnoy sfere [Security and Confrontation in the Information Domain]*, Moscow: RUSAVIA, 2010.

Armeiskii sbornik [Army Digest], Vol. 6, No. 7, 2000.

Badyakin, N. N., *Instruktsiya po ispol'zovaniyu shtabnoy matematicheskoy modeli «Predvideniye»: Uchebnoye posobiye [Manual on the Use of Headquarters Mathematical Model 'Foresight': Textbook]*, Moscow: Combined Arms Academy of the Armed Forces of the Russian Federation, 2001.

Barovska, A., ed., "Informatsiini vyklyky hibrydnoi viiny: Kontent, kanaly, mekhanizmy protydii' [Information Challenges of Hybrid Warfare: Content, Channels, Counteraction Mechanisms]," Kyiv: National Institute for Strategic Studies, 2016.

Barynkin, V. M., and M. A. Rodionov, *Ponyatiynyy apparat teorii informatsionnogo protivoborstva (Informatsionnoy bor'by) [Concepts of the Theory of Information Confrontation (Information Warfare)]*, Moscow: VAGSh, 1998.

Brammer, Y., and I. Pashchuk, *Tsifrovyye ustroystva [Digital Devices]*, Moscow: Publishing House "Higher School," 2004.

Dziuba, M., et al., *Narys teorii i praktyky informatsiino-psykholohichnykh operatsii [Essay on the Theory and Practice of Information and Psychological Operations]*, Kyiv: VITI NTUU "KPI," 2006.

Epple, H., "Rossiia v virtualnoi voine [Russia in a Virtual War]," *Vedomosti [Gazette]*, May 8, 2014.

Federal Law No. 24-F3, "Ob informatsii, informatizatsii i zashchite informatsii [About Information, Informatization, and the Protection of Information]," Moscow, February 20, 1995.

Gareev, M. A., "Voyna i sovremennoye mezhdunarodnoye protivoborstvo [War and the Modern International Confrontation]," *NVO*, No. 1, 1998.

Gareev, M. A., "O kharaktere vooruzhennoy bor'by budushchego [On the Nature of the Armed Combat of the Future]," *Vestnik Akademii voennykh nauk [Journal of the Academy of Military Sciences]*, No. 2, 2005, pp. 11–14.

Gareev, M. A., "Itogi deyatel'nosti Akademii voyennykh nauk za 2012 god i zadachi Akademii na 2013 god [Activities of the Academy of Military Sciences for 2012 and Academy Tasks for 2013]," *Vestnik Akademii voennykh nauk [Journal of the Academy of Military Sciences]*, Vol. 1, No. 42, 2013a, pp. 8–21.

Gareev, M. A., "Sistema znaniy o voyne i oborone strany na sovremennom etape [System of Knowledge about War and Defense of the Country in Modern Times]," *Vestnik Akademii voennykh nauk [Journal of the Academy of Military Sciences]*, Vol. 2, No. 43, 2013, pp. 7–14.

Gerasimov, Valery, "Po opytu Sirii [Syrian Experience]," *Voyenno-promyshlennyy kur'er [Military-Industrial Courier Online]*, March 7, 2016.

Government of the Russian Federation, "Federal Law No. 150, Ob Oruzhii [On Weapons]," December 13, 1996.

Government of the Russian Federation, "Federal Law No. 149, Ob informatsii, informatsionnykh tekhnologiyakh i o zashchite informatsii [On Information, Information Technologies and Information Protection]," Moscow, July 27, 2006.

Government of the Russian Federation, "Voyennaya doktrina Rossiyskoy Federatsii [Military Doctrine of the Russian Federation]," No. Pr-2976, December 25, 2014.

Government of the Russian Federation, "Doktrina informatsionnoy bezopasnosti Rossiyskoy Federatsii [Information Security Doctrine of the Russian Federation]," No. 646, December 5, 2016.

Grisé, Michelle, Alyssa Demus, Yuliya Shokh, Marta Kepe, Jonathan Welburn, and Khrystyna Holynska, *Rivalry in the Information Sphere: Russian Conceptions of Information Confrontation*, Santa Monica, Calif.: RAND Corporation, RR-A198-8, 2022.

Hai-Nyzhnyk, Pavlo, L. L. Zaliznyak, I. Y. Krasnodems'ka, Yu. S. Fihurnyy, O. A. Chyrkov, and L. V. Chupriy, *Ahresiia Rosii proty Ukrainy: Istorychni peredumovy ta suchasni vyklyky [Russia's Aggression Against Ukraine: Historical Background and Current Challenges]*, Kyiv: MP Lesya, 2016.

Hai-Nyzhnyk, Pavlo, *Rosiia proty Ukrainy (1990–2016 r.): Vid Polityky shantazhu i prymusu do viiny na pohlynannia ta sproby znyshchennia [Russia Against Ukraine (1990–2016): From a Policy of Blackmail and Coercion to a War for Absorption and Attempts to Destroy]*, Kyiv: MP Lesia, 2017.

Horbulin, Volodymyr, *Svitova hibrydna viina: Ukrainskyi Front [The World Hybrid War: Ukrainian Forefront]*, Kyiv: National Institute for Strategic Studies, 2017.

Iuzvyshyn, Ivan, *Informatsiologiya ili zakonomernosti informatsionnykh protsessov i tekhnologiy v mikro- i makromirakh vselennoy* [*Informationology or Patterns of Information Processes and Technologies in Micro- and Macroworlds of the Universe*], Moscow: Radio i sviaz, 1996.

Lata, V. F., V. A. Annenkov, and V. F. Moiseev, "Informatsionnoye protivoborstvo: Sistema terminov i opredeleniy [Information Confrontation: System of Terms and Definitions]," *Vestnik Akademii voennykh nauk [Journal of the Academy of Military Sciences]*, No. 2, 2019, pp. 128–138.

Libiki, M., "Shcho take informatsiina viina? [What Is Information Warfare?]," 2019.

Liddell-Hart, B. H., *Strategiya nepryamykh deystviy [Strategy of Indirect Actions]*, Moscow: Exmo, 2008.

Lytvynenko, Oleksandr, "Totalna viina po-Putinski: 'Hibrydna' viina RF proty Ukrainy [Total Putin-Style War: Russia's 'Hybrid' War Against Ukraine]," in *"Hibrydna" viina Rosiii—vyklyk i zahroza dlya Yevropy [Russia's Hybrid War—Challenge and Threat for Europe]*, Kyiv: Razumkov Centre, 2016.

Manoilo, A. V., *Gosudarstvennaya informatsionnaya politika v osobykh usloviyakh: Monografiya [Federal Information Policies Under Special Conditions: Monograph]*, Moscow: MIFI, 2003.

Manoilo, A. V., *Informatsionnyye voyny i psikhologicheskiye operatsii: Rukovodstvo k deystviyu [Information Wars and Psychological Operations: Call to Action]*, Moscow: Hotline—Telecom, 2018.

Matvienko, Yu., "'Tsvetniye' revoliutsii kak nevoyenniy sposob dostizheniya politicheskikh tseley v gibridnoy voyne: Sushnost', soderzhaniye, vozmozhniye mery zashity i protivodeystviya ['Color' Revolutions as Non-Military Means to Achieve Political Goals in 'Hybrid' War: Nature, Content, Possible Protection Measures and Countermeasures]," *Informatsionniye Voyny [Information Wars Journal]*, Vol. 4, No. 40, 2016, pp. 11–19.

Ministry of Defense of the Russian Federation, "Informatsionnaya voyna [Information War]," *Voyennyy entsiklopedicheskiy slovar' [Military Encyclopedic Dictionary]*, trans. Polina Kats-Kariyanakatte, Joe Cheravitch, and Clint Reach, webpage, undated-a. As of November 10, 2020: https://encyclopedia.mil.ru/encyclopedia/dictionary/details.htm?id=5211@morfDictionary

Ministry of Defense of the Russian Federation, "Informatsionnaya protivoborstvo [Information Confrontation]," *Voyennyy entsiklopedicheskiy slovar' [Military Encyclopedic Dictionary]*, undated-b. As of May 5, 2021: http://encyclopedia.mil.ru/encyclopedia/dictionary/details.htm?id=5221@morfDictionary

Ministry of Defense of the Russian Federation, *Kontseptual'nye vzglyady na deyatel'nost' Vooruzhennykh sil Rossiyskoy Federatsii v informatsionnom prostranstve [Conceptual Views on the Activities of the Armed Forces of the Russian Federation in the Information Space]*, 2011. As of July 8, 2020: http://ens.mil.ru/science/publications/more.htm?id=10845074@cmsArticle

Ministry of Defense of the Russian Federation, "Ministr oborony Sergey Shoygu nazval glavnoy tsel'yu informatsionnoy voyny Zapada protiv Rossii polnoye yemu podchineniye [Defense Minister Sergey Shoygu Called Complete Submission to the West as the Main Goal of the Information War of the West Against Russia]," webpage, June 26, 2019. As of November 13, 2020: https://function.mil.ru/news_page/country/more.htm?id=12238562@egNews

Noveyshiy slovar' inostrannykh slov i vyrazheniy [New Dictionary of Foreign Words and Expressions], Minsk: Contemporary Literary Man, 2006.

Novikov, V. K., *Informatsionnoye oruzhiye - oruzhiye sovremennykh i budushchikh voyn* [*Informational Warfare: The Warfare of Contemporary and Future Wars*], Moscow: Goriachaia Liniia—Telekom, 2011.

Novikov, V. K., "Nie ubit, no podchinit [Not to Kill, but to Subjugate]," *Voenno-promyshlennyi courier [Military-Industrial Courier]*, December 16, 2015.

Novikov, V. K., and S. V. Golubchikov, "Migrant istiny [Migrant of Truth]," *Voenno-promyshlennyi courier [Military-Industrial Courier]*, July 29, 2016.

Novikov, V. K., and S. V. Golubchikov, "Analiz informatsionnykh voyn za posledniye chetvert' veka [Analysis of Information Wars of the Last Quarter Century]," *Vestnik Akademii voennykh nauk [Journal of the Academy of Military Sciences]*, No. 3, 2017, pp. 10–16.

Nuzhdin, Y., "Informatsionniye voyny. Uroki devianostykh [Information Wars. Lessons of the Nineties]," *Flag Rodiny [Flag of the Motherland]*, November 22, 2000.

Orlansky, V. I. "Informatsionnoye oruzhiye i informatsionnaya bor'ba: Real'nost' i domysly [Information Weapons and Information Warfare: Reality and Speculation]," *Voennaya mysl' [Military Thought]*, No. 1, 2008, pp. 62–70.

Panarin, I. N., *Informatsionnaya voyna, PR i mirovaya politika [Information War, PR and Global Politics]*, Moscow: Hotline—Telecom, 2006.

Pievtsov, H., et al., *Informatsiino-psykholohichna borotba u voiennii sferi [Information and Psychological Warfare in the Military Domain]*, Kharkiv: Kharkiv National University, 2017.

Poteev, M., *Kontseptsii sovremennogo iestestvoznaniya. Uchebnik* [*Concepts of Modern Natural Science. A Textbook*], St. Petersburg: Piter, 1999.

President of Russia, *Voyennaya doktrina Rossiyskoy Federatsii* [*Military Doctrine of the Russian Federation*], Moscow, April 21, 2006.

Prokof'yev, V. F., *Taynoye oruzhiye informatsionnoy voyny: Ataka na podsoznaniye [Secret Weapon of Information Wars: Attacking the Subconscious]*, 2nd ed., Moscow: Sinteg, 2003.

Prysiazhniuk, D., "Zastosuvannya manipuliatyvnykh psykhotekhnolohii z boku Rosii v ZMI Ukrainy (Na prykladi Krymu) [Application of Manipulative Psychotechnologies by Russia in Ukrainian Media (on the Example of Crimea)]," *Visnyk Kyyivs'koho Natsional'noho Universytetu imeni Tarasa Shevchenka. Viys'kovo-spetsial'ni nauky [Journal of Taras Shevchenko National University of Kyiv. Military Sciences]*, No. 23, 2009, pp. 63–66.

Prysiazhniuk, M., "Osoblyvosti suchasnoho periodu informatsiyno-psykholohichnoho protyborstva [Peculiarities of the Modern Period of Informational-Psychological Confrontation]," in Y. Zharkov et al., eds., *Istoriia informatsiino-psykholohichnoho protyborstva [History of Information and Psychological Confrontation]*, Kyiv, Ukraine: Research and Publishing Department of the National Academy of Security Service of Ukraine, 2012.

Rodionov, M. A., "K voprosu o formakh vedeniya informatsionnoy bor'by [On the Question of the Ways of Waging Information Warfare]," *Voennaya mysl' [Military Thought]*, No. 2, 1998, pp. 67–70.

Sayfetdinov, K. I., "Informatsionnoye protivoborstvo v voyennoy sfere [Information Confrontation in the Military Sphere]," *Voennaya mysl' [Military Thought]*, No. 7, 2014, pp. 38–41.

Shevtsov, V. S., "Informatsionnoye protivoborstvo v globaliziruyushemsia mire: Aktual'nost', differentsiatsiya poniatiy, ugrozy politicheskoy stabil'nosti [Information Confrontation in a Globalizing World: Relevance, Differentiation of Concepts, Threats to Political Stability]," *University Bulletin [Vestnik universiteta]*, No. 5, 2015, pp. 206–211.

Slipchenko, V., "Novaya forma bor'by. V nastupivsheme veke rol' informatsii v beskontaktnykh voynakh budet lish' vozrastat' [A New Form of Combat. In the Coming Century the Role of Information in the Contactless Wars Will Only Increase]," *Armeiskii sbornik [Army Digest]*, No. 12, 2002, pp. 30–32.

Slipchenko, V., "Informatsionnyy resurs i informatsionnoye protivoborstvo [Information Resources and Information Confrontation]," *Armeiskii sbornik [Army Digest]*, No. 10, 2013, pp. 52–57.

Slovar' terminov i opredeleniy v oblasti informatsionnoy bezopasnosti [Dictionary of Terms and Definitions in the Area of Information Security], 2nd ed., Moscow: VAGSh, 2008.

Solomonov, Y. S., "Strategicheskaya tsel' [Strategic Goal]," Moscow: OOO Publishing House A4, 2014.

Soviet Military Encyclopedia, Moscow: Voenizdat, undated.

Sovietskaya voyennaya entsiklopediya [*Soviet Military Encyclopedia*], Vol. 8, Moscow: Voenizdat, 1980.

Stasiuk, V., *Psykholohichne zabezpechennia diialnosti viisk (syl)" [Psychological Support for the Activities of Troops (Forces)]*, Kyiv: Ivan Chernyakhovsky National University of Defense of Ukraine, 2014.

Svechin, A. A., *Strategiya v trudakh voyennykh klassikov [Strategy in the Works of Military Classics]*, Vol. 2, Moscow: Federal Military Edition, 1926.

Svechin, A. A., *Evolyutsiya voyennogo iskusstva [Evolution of Military Science]*, Moscow: Academic Project, 2002.

Tasbulatov, A., and V. Orlansky, "Razrabotka sovremennoy klassifikatsii vidov i sredstv porazheniya—neotlozhnaya zadacha voyennoy nauki [The Development of a Modern Classification of Types and Means of Destruction Is an Urgent Task of Military Science]," *Voennaya mysl' [Military Thought]*, No. 4, April 2007, pp. 55–61.

Trotsenko, K. A., "Informatsionnoye protivoborstvo v operativno-takticheskom zvene upravleniya [Information Confrontation on the Operational-Tactical Level]," *Voennaya mysl' [Military Thought]*, No. 8, 2016, pp. 20–25.

Tsygankov, V., and V. Lopatin, *Psikhotronnoye oruzhiye i bezopasnost' Rossii* [*Psychotronic Weapons and Security of Russia*], Moscow: Sinteg, 1999.

Turchenko, Fedir, and Halyna Turchenko, *Proiekt "Novorosiia" and novitnia Rossiisko-Ukrainska viyna* [*The Novorossiya Project and the Latest Russian-Ukrainian War*], Kyiv: Institute of History of Ukraine, 2015.

U.S. Army Field Manual 100-6, *Information Operations*, Washington, D.C.: Headquarters, Department of the Army, August 27, 1996.

Volkovskii, N. L., *Istoriya informatsionnykh voyn [History of Information Wars]*, Vol. 1, Saint Petersburg: OOO Polygon Publishing, 2003a.

Volkovskii, N. L., *Istoriya informatsionnykh voyn [History of Information Wars]*, Vol. 2, Saint Petersburg: OOO Polygon Publishing, 2003b.

Voloshyna, N., and M. Dziuba, "Vyroblennia u maibutnikh ofitseriv imunitetu proty nehatyvnoho informatsiino-psykholohichnoho volyvu [Development of Immunity of Future Officers to Negative Information and Psychological Influence]," *Visnyk Kyyivs'koho Natsional'noho Universytetu imeni Tarasa Shevchenka. Viys'kovo-spetsial'ni nauky [Journal of Taras Shevchenko National University of Kyiv]*, No. 30, 2013, pp. 34–37.

Vorontsova, L. V., and D. B. Frolov, *Istoriya i sovremennost' informatsionnogo protivoborstva [History and Modernity of Information Confrontation]*, Moscow: Hotline—Telecom, 2006.

Yavorska, H., "Hibrydna viina yak dyskursyvnyi konstrukt [Hybrid Warfare as a Discursive Construct]," *Stratehichni priorytet [Strategic Priorities]*, No. 4, 2016, pp. 41–48.

Yuzova, I., "Analiz Orhanizatsiyi ta vedennya informatsiyno-psykholohichnykh operatsiy pry vedenni hibrydnoyi viyny [Analysis of the Organization and Conduct of Informational-Psychological Operations in the Conduct of Hybrid Warfare]," *Zbirnyk naukovykh prats' Kharkivs'koho natsional'noho universytetu Povitryanykh Syl [Anthology of Research Works of Kharkiv National Air Force University]*, No. 2, 2020, pp. 40–44.

Zharkov, Y., et al., eds., *Istoriia informatsiino-psykholohichnoho protyborstva [History of Information and Psychological Confrontation]*, Kyiv, 2012.